BREAKING LIFE SCIENCES VOL I

THE GHOST GENOME

The Hidden Chemical Language Inside Us - and Why Medicine Has Barely Begun to Read It

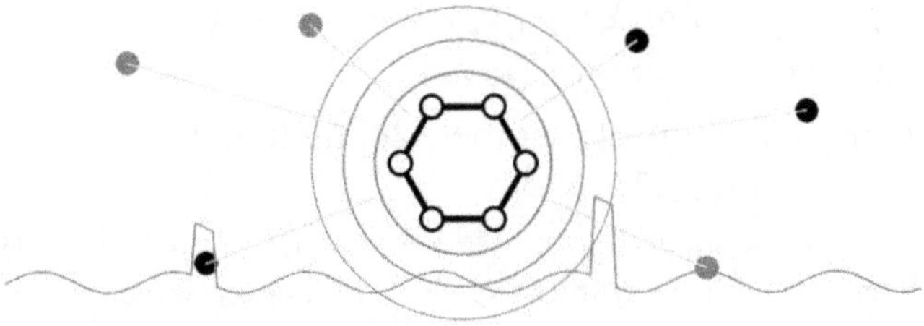

Ross O. Youngs

First Edition | September 2026

For permission requests or publisher information, contact:
Ross Youngs (Author/Publisher)
2545 Farmers Dr., Suite 370
Columbus, OH 43235
https://rossyoungs.com/
ISBN (hardcover): 979-8-9973605-2-8
ISBN (eBook): 979-8-9973605-3-5
Library of Congress Control Number: 2026920922
First Edition, September 2026

Printed in the United States of America

To the teachers who looked through the new
instrument
before the rest of us were ready.

And to the explorers who kept looking
when the system asked them to stop.

Contents

NARRATIVE ARCHITECTURE

Ten Stories, One Hidden World

Each story stands on its own and returns later with a different meaning.

1 **The Gene Sweepstakes** — *The Book of Life Was Too Short*

2 **The Doctor Who Drank the Evidence** — *When a Bacterium Rewrote a Disease*

3 **The Planet That Should Not Exist** — *A Modern Galileo Story*

4 **The Molecule in the Freezer** — *The Outlaw That Survived Its Corporate Death*

5 **The Twins Who Transferred a Metabolism** — *When a Community Changed the Host*

6 **The Viruses in the Mucus** — *The Immune Layer the Host Did Not Encode*

7 **The Lake That Would Not Fit in a Petri Dish** — *Scale Becomes an Instrument*

8 **The Chip That Returned the Microbe Home** — *When the Environment Became Part of the Experiment*

9 **The Day AI Solved the Right Problem** — *What AlphaFold Teaches - and What It Does Not*

10 **The Soil Microbe That Changed Global Health** — *From Natural Prior to Medicine*

VISUAL PROGRAM

List of Figures

AUTHOR'S NOTE

Four Levels of Certainty

This book makes a large claim: that modern biology has counted much of the hardware of human life while missing a vast portion of the chemistry that actually coordinates it. Some parts of that claim are already well established. Some are reasonable inferences from disconnected fields. Some are hypotheses born from direct observation and years of work in microbial natural-product discovery. They should not be blended together.

To keep the bargain with the reader, the manuscript uses four levels of certainty.

Established finding. These are observations supported by replicated experiments, accepted measurements, or strong bodies of peer-reviewed evidence. Examples include the near parity between human and bacterial cell counts in a reference adult; the influence of microbial metabolites on host receptors; the large fraction of mass-spectrometry features that remain unannotated; and the continuing importance of natural products and their derivatives in medicine.

Supported interpretation. These are conclusions that connect multiple established findings, but for which reasonable scientists may disagree about emphasis or causality. Examples include the argument that life science has overinvested in reductionist, single-target models, or that evolutionary chemical space is a richer starting point for many complex diseases than unconstrained synthetic space.

Author's hypothesis. These are propositions I believe are plausible and important enough to test, but which are not yet established. The two-billion-molecule estimate belongs here. So does the stronger version of the cryptic-assembly hypothesis: that host enzymes, microbial communities, dietary fragments, cell turnover, local metals, redox conditions, and transient reactions collectively generate a vast set of functional molecules that cannot be read directly from any one genome. In this manuscript, the term ghost molecule refers to an uncharacterized or poorly characterized small molecule, molecular family, or transient chemical state whose biological role is hidden by limits of access, concentration, timing, or annotation.

Commercial illustration. I have spent years trying to build tools that can reach this hidden chemistry. Those efforts shape my point of view. When I discuss Biosortia or Industrial-Scale Microbiome Mining, I am describing one proposed route into the problem, not claiming that a company has already proved the full thesis of this book. Publicly reported company work is identified as such and is separated from peer-reviewed evidence.

A good investigation does not hide uncertainty. It gives uncertainty a name, places it in the open, and then asks what experiment would reduce it.

Personal scenes in this edition are reconstructed from public interviews, company history, and the author's recollection. Unless quotation marks are used, dialogue is paraphrased. Exact operational details are intentionally limited where disclosure would compromise trade secrets, intellectual property, or source protection.

Three Working Definitions

The central terms in this book are meant to make an unseen problem discussable, not to disguise uncertainty with a dramatic name.

The ghost genome is my working name for the dynamic chemical execution layer produced when human genes, microbial genes, viruses, food, environment, anatomy, and time interact. It is not a hidden chromosome and not a replacement for genomics. It is the changing output of the whole living system - the part that cannot be read directly from any one sequence.

A ghost molecule is a physical chemical entity, molecular family, conjugate, complex, or transient state whose identity or role is hidden by limits of access, abundance, timing, stability, location, or annotation. Some ghost molecules will prove new. Others will be known chemistry doing unrecognized work in an unexpected place or combination.

A ghost activity is a reproducible biological effect for which the responsible chemistry has not yet been resolved. An active fraction, a community phenotype, or a time-limited signal can be real before the exact molecule is known. The scientific obligation is to move from activity to identity and from identity to causality without pretending that one stage has already completed the next.

These definitions create a progression. First we see a trace. Then we show that the trace or activity repeats. Then we preserve its context, recover enough material, identify the chemistry, and test whether it actually changes biology at plausible exposure. The word ghost should disappear as knowledge improves.

What I Believe

I believe the evidence already justifies a coordinated effort to observe this chemical layer much more deliberately than we do today.

That belief is stronger than saying the idea is merely interesting and weaker than claiming the full thesis has been proved. We already know that microbes transform food, bile acids, medicines, and host products; that microbial molecules can reach human tissues and receptors; that many measured chemical features remain unresolved; and that location and timing can determine function. What remains uncertain is the scale, organization, and therapeutic yield of the hidden layer.

Vision matters at this stage because instruments are never built in a vacuum. Someone must believe that an unobservable world is worth making observable before the first decisive dataset exists. The responsible form of belief does not lower the standard of proof. It raises the seriousness of the experiment.

The mission is therefore practical: turn hidden activity into reproducible chemistry, reproducible chemistry into causal knowledge, and causal knowledge into interventions that improve human and planetary life. Therapeutics are the nearest and most demanding test. Longevity, agriculture, ecosystem resilience, and closed-loop human habitats are later horizons, not promises attached to the first result.

The book will keep returning to one discipline: confidence should be proportional to evidence, while ambition should be proportional to the importance of the unanswered question.

PREFACE

Permission to Look

The first scientific instruments I learned to respect were not in a university laboratory. They were medical-imaging systems that my father installed, and from the age of sixteen through college I helped wherever an extra pair of hands was useful.

A scanner is a promise made physical. Components have to be aligned. Cables have to reach the right place. Power, shielding, signal, software, calibration, and the anatomy itself all have to agree. When the system works, something invisible a moment earlier appears on a screen. A shadow becomes an organ. A hidden injury becomes a finding. Long before I had language for it, I learned that discovery depends on the instrument that makes the hidden thing measurable.

The lesson followed me into every industry I entered.

At the Florida Institute of Technology I studied environmental science. At Purdue I moved into industrial engineering. The two fields gave me different vocabularies for the same obsession: resources. Environmental science asks how we avoid wasting the natural world. Industrial engineering asks how we remove waste from a process. I became interested in the point where those questions meet - how to make a system better, faster, less expensive, and meaningfully different without pretending that physics can be negotiated.

Manufacturing taught me another lesson: an accepted limit is often the memory of an old process.

My companies redesigned products, materials, and production systems because the conventional answer was frequently built around equipment, assumptions, and economics that no one had recently challenged. That approach brought success. It also brought the opposite. One of the enterprises I built had reached the Inc. 500 five years in a row. Then the underlying technology changed. A market that had seemed durable disappeared with astonishing speed. In my public telling of the experience, sales in that technology went from essentially all of our business to nothing almost overnight.

Survival required more than improving the product that yesterday's customers had wanted. It required accepting that the architecture had changed.

Years later, when I began looking closely at the economics of pharmaceutical research, I recognized the same danger. An industry can become exceptionally skilled at optimizing a system while the world around the system changes. Each individual decision can be rational. The combined result can still move backward.

The path from manufacturing to microbiome science did not begin with a drug target. It began with algae, water, and a mass-balance problem.

Algae are easy to admire in a photograph and difficult to recover from a lake, pond, or reactor. The useful biomass is microscopic and dilute. Most of what a pump moves is water. Membranes foul. Pressure rises. Energy costs swallow the value of the material. The bottleneck was not the promise of biology. It was physical access.

The engineering question was straightforward: how do you recover very small living solids from very large volumes of water quickly, gently, and economically? The work eventually produced a separation platform and, in 2009, a $6 million ARPA-E research program around algal harvesting. Then the meaning of the platform changed.

If we could recover microbial communities at scale, perhaps scale was not merely a manufacturing step after discovery. Perhaps scale was part of the scientific instrument itself.

A molecule present at one part per billion is almost nothing in a milliliter. Across millions of liters, it is matter. Once it becomes matter, it can become a fraction. A fraction can become repeated measurements. Repeated measurements can become a structure, a synthesis, an assay, and an argument that another laboratory can challenge.

That realization led to an impolite question that has followed me through every chapter of this book:

What if the bottleneck everyone accepts is not a law of nature, but a consequence of how the problem has been framed?

The most dangerous sentence in innovation is not *That is impossible.* It is *We already know how this works.*

I have heard versions of that sentence throughout my career. Sometimes it came from an engineer who could explain why a material could never be manufactured. Sometimes from a financier who knew why a market would never form. Sometimes from a scientist who had spent decades mastering a field and could list, with admirable precision, every reason a different approach was unlikely to succeed.

Expertise matters. Civilization depends on it. But expertise carries a hidden cost: once we become fluent in a system, we begin to confuse the system's map with the territory itself.

I did not enter microbiome science as a career academic trained to defend one narrow mechanism. I entered as an inventor and entrepreneur who had already learned that a new instrument can turn an accepted limitation into an engineering variable. I also entered with an unusual advantage and a corresponding responsibility. As a generalist, I could move among environmental science, engineering, manufacturing, natural-products chemistry, data, business, and drug discovery. I could ask naive questions without risking a career built around one answer. But I also had to depend on specialists, invite correction, and resist the entrepreneur's temptation to make a compelling story outrun the evidence.

That bargain with the reader is essential here.

This is not the friendly cartoon version of the microbiome that appears in advertisements for yogurt and supplements. It is the overwhelming chemical reality beneath it: communities too complex to culture, signals too dilute to isolate, metabolites too transient to catch, and molecular interactions occurring in combinations that reductionist experiments routinely dismantle before measurement.

For more than a decade, I have watched extraordinary scientists make discoveries that should have forced a broader reconsideration of human biology. Microbes shape immune development. They transform bile acids, amino acids, dietary compounds, and drugs. They make ligands for human receptors. They influence barrier integrity and metabolic physiology. Natural products repeatedly supply structures that human medicinal chemistry did not imagine on its own. Mass spectrometers detect oceans of molecular features that libraries cannot name.

Each fact is acknowledged. The larger implication is often avoided.

The implication is that the human organism is not fully described by the human genome, the human proteome, or even the catalog of microbial genes living beside us. We are a continuously operating chemical ecosystem. Human cells, microbial cells, viruses, food, environmental inputs, and the products of cellular death and renewal are joined in a moving network of transformation. The molecules created in that network are not merely waste. Some are instructions. Some are negotiations. Some are alarms, brakes, invitations, border controls, and repair signals.

And most remain unnamed.

The purpose of this book is not to announce that every unknown mass-spectrometry peak is a medicine, that microbes are always benevolent, that artificial intelligence is overhyped, or that synthetic chemistry has failed. Those claims would be easy to attack because they would be wrong. The purpose is to investigate a narrower and more consequential possibility: that life science has undercounted the functional chemical layer of biology, and that this undercount helps explain why drug discovery becomes more expensive even as our tools become more powerful.

To tell that investigation as a story, I have built the book around ten episodes. At first they may seem only loosely connected: a betting book passed around a bar at Cold Spring Harbor; a physician drinking a bacterial culture; a planet orbiting where no giant planet was supposed to be; a microorganism hidden in a household freezer; twins whose microbiomes changed the metabolism of mice; viruses caught in mucus; a lake that would not fit in a Petri dish; a plastic chip returned to soil; an artificial-intelligence system that solved the right kind of problem; and a soil microbe whose chemistry reached hundreds of millions of people.

The connection is not that every outsider was right. It is not that every natural product becomes a medicine. It is not that new instruments abolish the need for skeptical review. The connection is that reality repeatedly exceeds the categories we built to contain it. The most important discoveries often arrive first as an observation that the old map has nowhere to place.

The system is the antagonist in this story, but scientists are not the villains. The system is a set of incentives that rewards what can be reviewed, reproduced, patented, published, and advanced through familiar gates. It is excellent at improving known engines. It is much less comfortable funding a new vehicle whose dashboard has not yet been invented.

This book asks us to widen the map without lowering the standard of proof.

The Mission Before the Medicine

The mission of The Ghost Genome is not to prove one company right. It is to make a missing category of biology visible enough that many laboratories, institutions, and nations can test it.

The work begins with an observatory, not a product. We need ways to access living chemistry without erasing the ecology that produced it; preserve place and time; separate complex mixtures; measure beyond existing libraries; connect unknown features to activity; confirm structures; and feed every success and failure back into the next experiment.

That observatory will use genomics, but it will not stop at genomics. Sequence is the durable source code and architectural language of the system. RNA and proteins are loaded programs and machinery. Metabolites and molecular assemblies are messages, controls, inputs, outputs, and changing system state. Microbes, food, medicines, viruses, and environment provide additional code paths and external inputs. No single layer is the operator. The operator is the living interaction among them.

Artificial intelligence can help us organize this complexity, but only after biology has been allowed to create a training set that includes what our older methods overlooked. The missing information is not merely hidden in a larger database. Some of it has not yet been physically recovered, preserved, identified, or connected to function.

This book stands on its own by answering the first question: What are we missing, and how would we know it was there? A later question is how innovators and institutions should move when the evidence points beyond the accepted map. A still deeper question is how a civilization might use continuous chemical state information responsibly. Those questions belong to a larger arc, but the intellectual burden here is complete: establish why the hidden layer is real enough, consequential enough, and observable enough to deserve a serious field.

The promise is not that the ghost genome contains a cure for everything. The promise is that exploring it can expand the starting material from which better questions, better models, and eventually better interventions are built.

PROLOGUE

The Backward Law

If generation becomes nearly free, biological relevance becomes the scarce resource.

The Backward Law

The first thing a high-resolution mass spectrometer gives you is not an answer. It is a field of signals.

A cyanobacterial bloom has already done the chemistry before the sample reaches the instrument. Sunlight, nutrients, temperature, predators, viruses, competition, and time have shaped a living community. Collection preserves only a moment. Extraction translates that moment into fractions. Chromatography separates part of the crowd. Then the instrument draws peaks across a screen - some familiar, many not.

A peak can look decisive. It has a mass, an intensity, a retention time, and sometimes a fragmentation pattern. Yet it may represent an isotope, an adduct, a contaminant, a fragment, a known molecule in an unfamiliar form, or a structure no one has previously reported. The instrument does not tell you which story is true. It tells you that something physical passed through the system and left a trace.

In a public 2026 collaboration involving a prefractionated library from a harmful algal bloom in Muskegon, Michigan, researchers isolated previously unreported micropeptins, confirmed another cyanopeptide, and used product-ion queries and molecular networking to recognize related cyanopeptide classes across the collection. The work did not announce a medicine. It demonstrated something more basic and, for discovery, more important: an intact community could contain a broader chemical vocabulary than the handful of molecules routine monitoring was designed to find.

Every peak demanded restraint. A feature is not necessarily a unique molecule. A molecule is not necessarily biologically important. Activity in an assay is not a therapy. Yet the physical material existed before the model, before the database match, and before anyone knew what question to ask of it. The sample was not a computational possibility. It was evidence that biology had already written more chemistry than our catalog could read.

That is the scene behind this book: better instruments revealing more chemical reality at the same moment the economics of drug discovery appear to be moving backward.

I had seen a version of that paradox before.

In the optical and video-disc industry, we worked inside a technology that seemed solid enough to build a company around. We improved products and processes. We won customers. We grew. The organization learned what quality meant, what buyers valued, and how to execute reliably. Then the architecture changed. The market did not send a courteous notice explaining that the metrics of success were about to expire. It simply moved.

Going from nearly all of a company's sales in a technology to almost none is not an intellectual exercise. It is payroll, leases, suppliers, equipment, families, pride, and the sickening realization that yesterday's expertise can become today's fixed cost. The experience left me with a permanent suspicion of systems that congratulate themselves for improving while their underlying productivity deteriorates.

Pharmaceutical research presents a larger and more consequential version of the problem.

If a car company spent eighty times more money to produce the same number of working cars, the board would not call it progress. If each new generation of factories became more automated, more computationally sophisticated, and more expensive - while the number of vehicles that survived the road test continued to fall - no one would blame the road for being complicated.

Yet that is close to the economic mystery at the center of modern drug discovery.

In 2012, Jack Scannell and his colleagues gave the mystery a memorable name: Eroom's Law, Moore's Law spelled backward. Moore's Law had become shorthand for the astonishing improvement of computing power. Eroom's Law described the opposite trajectory in pharmaceuticals. From the middle of the twentieth century onward, the number of new drugs approved per inflation-adjusted billion dollars of research and development spending fell dramatically - roughly halving every nine years in the historical analysis.

The decline did not happen because laboratories became primitive. The industry acquired robotic screening, combinatorial chemistry, genomic databases, structural biology, transgenic animals, cloud computing, machine learning, and now generative artificial intelligence. At every step, the tools improved. The output per dollar deteriorated.

That does not prove a single cause. Regulation became more demanding. Trials grew larger. Success in one disease left harder diseases behind. Companies became more cautious after spectacular failures. Biology resisted our clean diagrams. Scannell and his coauthors offered several explanations, including the tendency to overestimate the value of new technologies and the problem of discovering "better than the Beatles" drugs after earlier medicines had already captured the easiest gains.

Two Curves Moving in Opposite Directions

Moore's Law and Eroom's Law form the visual paradox of modern drug discovery.

Computing made calculation, storage, simulation, and generation faster and cheaper. Pharmaceutical research gained those benefits, yet the historical output of approved medicines per inflation-adjusted research dollar moved in the opposite direction. The two curves do not prove that computers failed or that biology became more difficult on a schedule. They expose a mismatch between the rate at which we can create possibilities and the rate at which living systems can validate them.

AI inherits the favorable curve. It can generate structures, hypotheses, targets, and experimental plans at a speed no human team can match. Biology still charges the physical cost of being wrong. Molecules must be made, measured, delivered, metabolized, tolerated, and tested in systems that did not evolve to honor a prediction.

As generation becomes cheaper, a poorly grounded search can produce more false confidence rather than more medicines. The scarce resource becomes not imagination but biological relevance: starting points and experiments with a higher probability of surviving contact with life.

This is why the next advance cannot be only a faster search engine. It must also be a better way to decide where to search and a better observatory for learning what biology has already made.

But the backward law raises a question that cannot be dismissed with a list of costs:

What if the industry has been searching increasingly well inside a space with a declining base rate of success?

Imagine looking for a habitable island by widening the ocean. The radar becomes faster. The ship becomes larger. The maps become more detailed. Yet each new square mile of water contains less land than the last. Throughput rises while the chance that any one observation matters falls.

That is the intuition behind what I call the dilution problem. A viable drug must satisfy several constraints at once. It must influence the right biology. It must reach the right tissue. It must remain stable long enough, but not forever. It must avoid dangerous ion channels, organs, immune reactions, metabolites, and drug interactions. It must be manufacturable, formulatable, and clinically useful at a dose patients can tolerate. A compound that excels at one dimension can fail at any other.

If the design universe expands much faster than the prevalence of compounds satisfying all those constraints, then more searching does not automatically mean more success. It can mean more convincing false positives.

This is why the phrase *AI will search chemical space* deserves a second question: which chemical space, and with what prior probability that the structures inside it have ever functioned in a living system?

The conventional answer begins with what can be enumerated or synthesized. Vendor libraries contain billions of make-on-demand structures. Virtual spaces are vastly larger. Generative systems can propose molecules faster than chemistry teams can test them. These capabilities are real and useful. But speed cannot repair a poor starting distribution. A model trained primarily on the artifacts of prior medicinal chemistry may learn the strengths and blind spots of that chemistry with equal enthusiasm.

Nature begins elsewhere.

Every natural product exists because a living system made it. That statement does not guarantee that it is safe, useful to humans, or even stable outside its native niche. Nature evolves poisons as readily as medicines. But a physically recovered natural molecule has cleared a threshold that a virtual structure has not: it has been synthesized by biology, persisted in a chemical environment, and participated - somehow - in a living ecology.

That is not proof of drug viability. It is a prior.

And priors matter when the search space is enormous.

The natural-product record has been trying to tell us this for decades. David Newman and Gordon Cragg repeatedly cataloged the origins of approved medicines and showed that natural products, their derivatives, and structures inspired by them remain deeply represented across drug discovery, especially in oncology and infectious disease. The lesson is not that synthetic chemistry is irrelevant. The lesson is that synthetic chemistry often performs best when nature has already supplied a biologically meaningful starting point.

Still, even *natural products* describes only the chemistry we have managed to collect, culture, isolate, and identify. That is a thin shoreline compared with the ocean beyond it.

The Observability Gap

A molecule can exist and still be absent from science.

It may be produced only when several organisms compete. It may exist in one layer of mucus for seconds. It may bind a membrane, metal, protein, or neighboring molecule and disappear when the sample is diluted. It may be present at a concentration sufficient for a receptor beside the producing cell and insufficient for detection in blood. It may fragment in a way no library recognizes. It may be visible as activity while its exact identity remains unresolved.

Each failure of access creates a failure of training data. AI cannot learn the structure-function grammar of molecules that were never sampled, preserved, separated, identified, or linked to outcomes. A larger model trained on the same observable slice can become more fluent in the slice without revealing the missing world.

The observability gap is therefore not a minor analytical inconvenience. It is a boundary on what biology can teach computation, what medicine can treat as a candidate mechanism, and what institutions can recognize as an investable starting point.

Closing the gap requires intent. We must design the collection, preservation, measurement, and validation around the possibility that the important object is rare, local, conditional, or relational. The unobservable becomes observable only when the instrument is built for the way the phenomenon actually exists.

Most environmental microbes do not grow under standard laboratory conditions. Many microbial genes are silent outside their communities. Low-abundance signals disappear in small samples. Molecules change as neighboring species, host enzymes, diet, oxygen, metals, and pH change. A gene cluster may reveal a biosynthetic capacity without revealing the final molecule that exists in a tissue at a particular moment. An untargeted mass spectrometer may detect the event but fail to name it.

So the real mystery is larger than Eroom's Law. It is an accounting failure.

We counted about twenty thousand human protein-coding genes and called the genome a book of life. We later cataloged millions upon millions of microbial genes associated with human populations and called the microbiome a second genome. Both achievements were monumental. But genes are not the full execution of biology. They are capacities and instructions. The living system is what happens when those instructions meet context.

Context produces chemistry.

A reference adult contains on the order of thirty trillion human cells and a comparable or slightly larger number of bacterial cells, with enormous variation by person, body site, and circumstance. Those cells continuously consume, secrete, transform, exchange, and recycle molecules. Food adds a changing external library. The liver and intestine circulate bile acids through microbial transformations and back again. Dietary tryptophan becomes microbial indoles. Plant polyphenols become smaller microbial metabolites. Drugs can be activated, inactivated, or toxified by the microbiome. Phages restructure bacterial populations and move genes. Dying cells become feedstock for living ones.

There is no central control room.

And yet the system usually holds temperature, pH, barrier function, immune tolerance, energy balance, and tissue repair within ranges compatible with life.

What chemical language makes that possible?

How much of it have we never measured?

What if many chronic diseases are not failures of a single target, but failures of coordination across that language?

And what if the next great source of medicine is not an infinite catalog of molecules we can imagine, but the finite, physical, evolution-tested chemistry we have barely learned to see?

Those are the questions behind *The Ghost Genome.*

The phrase is deliberately provocative. The genome is not a molecule catalog, and the molecules described in this book are not ghosts because they are supernatural. They are ghosts because they leave traces without revealing themselves. We see an unexplained biological effect, a spectral feature without a library match, a disease association without a mechanism, or a receptor response without a known ligand. Something moved through the room. The instrument registered it. The name is missing.

The investigation begins in a bar at Cold Spring Harbor Laboratory, where some of the world's smartest geneticists placed money behind a question they believed the Human Genome Project was about to answer.

They were not merely wrong about the number.

They were wrong about what the number could explain.

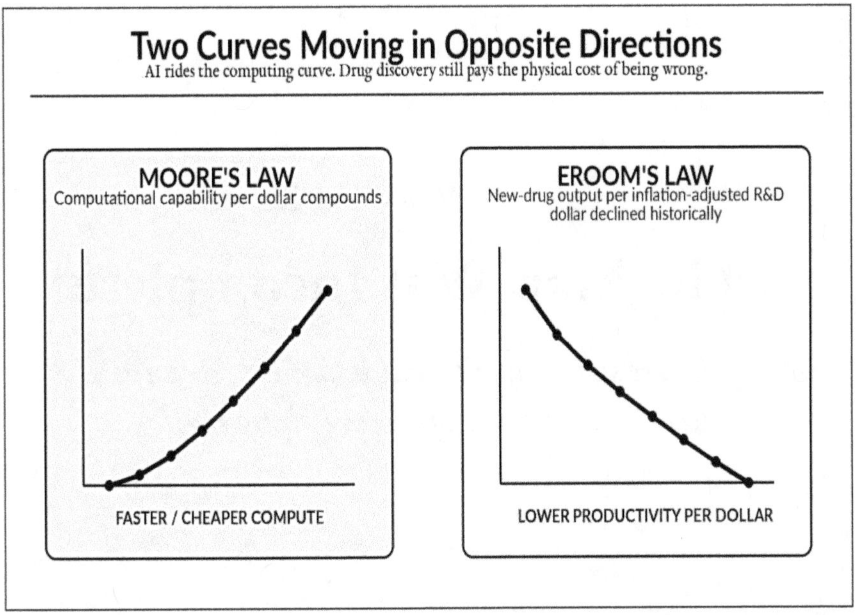

Figure 1. Two curves moving in opposite directions: computational capability compounds while historical pharmaceutical R&D productivity declines. Conceptual comparison; not a current forecast.

PART I

The Map Was Incomplete

Three stories about what happened when the accepted inventory stopped matching reality.

• • •

CHAPTER 1

The Gene Sweepstakes

The Book of Life Was Too Short

STORY 1

THE GENE SWEEPSTAKES

The Book of Life Was Too Short

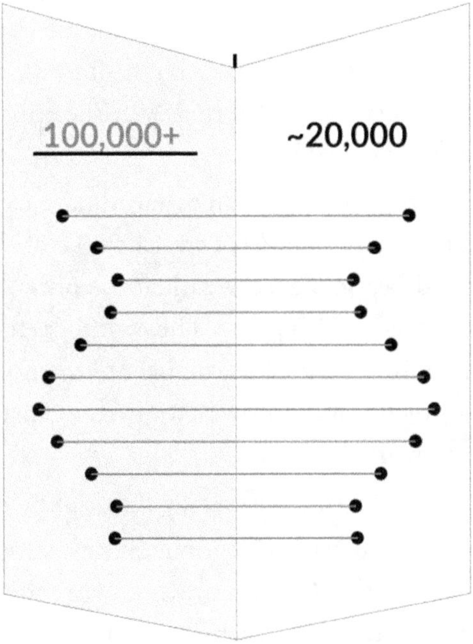

THE TURNING POINT

The parts list was real. The missing complexity lived in relationships.

In May 2000, some of the people closest to the Human Genome Project gathered at Cold Spring Harbor Laboratory on Long Island. The annual genome meeting had the atmosphere of a field nearing a summit. Machines were producing sequence faster. International teams were racing toward a draft. The language surrounding the project had grown correspondingly large: humanity was reading the Book of Life.

Then, in a bar, the scientists began betting on how many words the book contained.

The question was not framed as philosophy. It was a number: How many protein-coding genes does a human being have?

Ewan Birney, a British bioinformatician involved with the Ensembl genome-annotation project, turned the disagreement into a friendly wager. He carried a betting book. He also carried a plastic cup to collect money. Rules were written down, followed by footnotes, followed by more footnotes as participants discovered that even the apparently simple word *gene* became slippery at the edges.

The predictions sprawled across an extraordinary range. Some estimates were near 20,000. Others climbed toward 200,000. When the Human Genome Project had begun a decade earlier, the prevailing public estimate had commonly exceeded 100,000. The number felt proportional to human exceptionalism. Surely language, memory, consciousness, immunity, development, and culture required a genetic inventory much larger than that of a worm or a plant.

Then comparative data began lowering the ceiling.

A result from the pufferfish genome suggested something closer to 27,000. According to the history preserved by the National Human Genome Research Institute, the reaction was not serene recalibration. One version of the response was essentially: this person must be mad; there cannot be only 27,000.

But the number kept falling.

More than 460 participants eventually entered GeneSweep. The winning prediction, submitted by genome scientist Lee Rowen, was 25,947. The assessed count at the time of the contest's close was 24,847.

Modern estimates are nearer twenty thousand, depending on annotation rules and what exactly is counted.

The comedy of the betting book should not obscure the accomplishment around it. The Human Genome Project was one of the most consequential scientific achievements in history. It created a reference sequence, technologies, standards, databases, and a culture of open international collaboration that transformed nearly every branch of biomedicine.

The surprise was not that the project failed.

The surprise was that it succeeded and still made the old metaphor look inadequate.

The genome was called the Book of Life. Political leaders predicted a new era in which diseases could be read directly from DNA and corrected with precision. Popular accounts often treated the sequence as though humanity had finally opened the instruction manual to itself.

Then the instruction manual contained fewer conventional instructions than expected.

A human being did not become less complex because the count was lower. The count revealed that gene number was a poor proxy for complexity. Alternative splicing, noncoding regulation, epigenetic state, protein modification, cellular architecture, developmental timing, feedback, and network behavior multiplied what a modest inventory could do. The genome was not a short book. It was a compressed language whose meaning depended on context.

GeneSweep offers the first clue to the ghost genome because it exposed a recurring error in science: we count what the instrument can count, then allow the count to define what matters.

The error is understandable. Durable objects are easier to inventory.

DNA is stable enough to sequence. RNA can be captured. Proteins can be separated and measured. Cells can be sorted. These are the relatively persistent layers of biological hardware.

Small molecules are less cooperative. They span an enormous range of polarity, size, concentration, stability, and lifetime. Some exist for seconds. Some appear only at a tissue boundary. Some are consumed as quickly as they are made. Some ionize beautifully in one instrument and nearly disappear in another. Some share a mass but not a structure. Some produce fragments that resemble known compounds without matching any reference standard.

A gene can sit in a database for decades. A signaling molecule can vanish while the sample is being carried down the hall.

A wager about definitions

GeneSweep was entertaining because it looked precise. It was also a warning about the way precision can conceal a moving target.

What counted as a gene? Did a protein-coding sequence count once if it produced several splice forms? How should pseudogenes be handled? What about overlapping transcripts, short open reading frames, and sequences whose function was uncertain? The wager required rules because the object being counted was not as clean as the question implied.

That complication did not make the genome meaningless. It made the genome biological.

In engineering, a count is useful only when the unit remains stable. A factory may say it produced one thousand parts, but the number says very little if half the parts are assemblies, some are reworked components, and the quality standard changed halfway through the shift. The unit has to carry the meaning.

Biology repeatedly confronts the same problem. We speak of a species, a cell, a gene, a metabolite, or a disease as though the boundary were self-evident. Then a new instrument reveals subtypes, isoforms, states, dependencies, and relationships that do not fit the old bin.

The GeneSweep betting book became a physical record of that tension. The handwritten estimates represented not only different calculations but different mental models of where complexity lived. A high estimate implied that the human organism required a huge catalog of distinct protein instructions. A lower estimate implied that complexity had to emerge elsewhere - in regulation, combinations, timing, structure, and interaction.

The winning number therefore did not merely settle a pool. It changed the question from *How many instructions make a human?* to *How can so few durable instructions generate so many changing states?*

That second question leads directly toward the ghost genome.

If the biological unit is a dynamic chemical state rather than one gene or one stable molecule, then counting becomes harder. A molecule can be present for seconds. It can exist only at a boundary. It can be produced by one organism and modified by another. It can be inactive in blood and decisive in mucus. Its meaning can depend on concentration, partner, conformation, and timing.

The answer may still be countable. But the counting rules have to follow the biology rather than forcing the biology into the easiest ledger.

The ceremony and the metaphor

The public life of the Human Genome Project unfolded in two climactic moments. In June 2000, leaders from the public consortium and Celera Genomics stood with political leaders to announce working drafts. In 2003, the project celebrated the completion of the reference sequence near the fiftieth anniversary of the Watson and Crick paper on DNA structure.

The ceremonial language was understandable. Scientific projects need metaphors large enough to explain why society should care. A sequence of billions of chemical letters is difficult to translate into public meaning. *The Book of Life* turned an abstract infrastructure into a civilizational event.

The metaphor also created expectations that no responsible genome scientist had promised in such simple form. If DNA was the book, disease looked like a typographical error. Read the passage, identify the mutation, correct the letter, and restore the person. For some disorders, especially those driven by a single gene with a strong effect, that logic has produced extraordinary progress. Genetic diagnosis, targeted therapies, and gene-based medicines have changed lives.

For common disease, the text behaved less like an instruction manual and more like a set of conditional possibilities. The same variant can matter differently across ancestry, diet, infection history, age, sex, medication, development, and environment. Thousands of variants can each contribute a small shift. The relevant tissue may change before symptoms appear. Cells sharing the same genome can become neurons, hepatocytes, immune cells, and epithelium because they interpret different portions of the code in different states.

The genome was not wrong. The metaphor was too static.

A piano score contains the notes without containing one definitive performance. Tempo, phrasing, instrument, room, conductor, and musicians shape the sound. A genome contains durable sequence without containing every molecular state produced by a lifetime of development and interaction.

The lower gene count made this distinction unavoidable. Humans did not require ten times as many protein-coding genes as simpler organisms because biological complexity can be generated through reuse, regulation, combination, and context.

One gene can produce multiple RNA transcripts. One protein can be modified after translation in many ways. The same signaling component can participate in different pathways depending on cell type and timing. Enhancers can act across long genomic distances. Chromatin state changes which regions are accessible. Mechanical forces and metabolites alter transcription. Protein complexes assemble and dissolve. Feedback loops create memory.

The durable instruction is only the beginning of the execution.

This matters to the ghost-genome thesis because the missing chemical layer is not an optional decorative output. Metabolites participate in the regulation that gives the genome meaning. Acetyl-CoA, S-adenosylmethionine, alpha-ketoglutarate, NAD+, oxygen, iron, and other molecules influence chromatin and enzyme activity. Diet and microbes change the availability of some of those molecules. The chemistry does not merely occur after gene expression. It can change which genes are expressed next.

The arrows run both ways at the level of regulation.

DNA does not receive amino-acid sequence information back from a protein, and the central dogma remains correct within its intended domain. But cells continuously use chemical state to decide how the genome will be read. A parts list becomes behavior through feedback.

GeneSweep therefore revealed more than a surprisingly small number. It exposed the poverty of a one-directional public imagination.

The first expansion of the book

One context remained largely offstage when the human genome was being completed: the microbial world living in and on the body.

At the time, most medical education still presented microbes through the lens of infection. The intellectual inheritance ran through germ theory and the great victories against pathogens: identify the organism, establish causation, eliminate the invader. That framework saved uncounted lives. It enabled sanitation, vaccination, antibiotics, and modern infection control.

But a tool designed to identify invaders can make residents look unimportant.

The bacteria of the gut were commonly described as *flora*, a vague garden of passengers that helped digest food and occasionally caused trouble. Standard culture methods favored the minority of organisms willing to tolerate laboratory oxygen, nutrients, isolation, temperature, and timing. What could not be grown could not easily be studied. What could not be studied was demoted to background.

Then sequencing removed the requirement that a microbe first cooperate with a Petri dish.

The Human Microbiome Project and related international efforts helped turn a niche field into an expanding map of microbial communities. Early gut catalogs reported millions of nonredundant microbial genes. Later global catalogs expanded to hundreds of thousands of reconstructed genomes and vast predicted protein collections across human populations.

These numbers require care. A person does not carry every gene in a global catalog, just as one library patron does not possess every book in the national archive. But the comparison revealed the scale of collective microbial capacity associated with our species. The human genome was no longer the only relevant instruction set.

The first accounting error had been to assume that human complexity would be proportional to the number of human genes.

The second was to assume that adding microbial genes would complete the account.

A Species Name Is Not a Factory Specification

A microbial species name can sound like a standardized model number. It is not. In bacteria, organisms assigned to the same species can share highly similar DNA across the portions of their genomes they have in common while carrying substantially different sets of accessory genes. Those additional genes may arrive or disappear through plasmids, phages, genomic islands, transposons, and other forms of horizontal transfer. They can change which nutrients a strain can consume, which drugs it can transform, which competitors it can resist, and which specialized molecules it can produce. Two people may therefore carry the same named bacterial species while hosting different chemical factories.

Fungi make the lesson even broader. The approximately 1,800 *Saccharomyces* strains sometimes cited represent strains distributed across the yeast genus, not 1,800 versions of one species. Yet even within the single species *Saccharomyces cerevisiae*, a survey of 1,011 isolates found 4,940 coding units shared by every isolate and another 2,856 found only in subsets of the population. Yeast strains can also differ in chromosome-set number, extra or missing chromosomes, gene copy number, and hybrid ancestry. The variable portion of the yeast pangenome is especially enriched in functions involving stress, cell-to-cell interaction, and secondary metabolism—the functions most likely to matter when an organism confronts a changing ecology.

Filamentous fungi make the chemical consequence particularly clear. Some strains carry accessory chromosomes or biosynthetic gene clusters that other members of the same species lack. In one fungal species, approximately 40 percent of identified secondary-metabolite clusters varied in presence or absence among nineteen isolates. One strain may therefore possess the machinery to make a peptide, polyketide, terpene, toxin, signal, or competitive molecule that another strain bearing the same species name cannot make at all. Even possession does not guarantee production: a fungal pathway may remain silent until a competitor, host, nutrient, temperature shift, or chemical stress activates it.

These examples prevent us from mistaking a taxonomic label for a functional description. But even the exact strain genome remains only a catalog of capacity. It cannot tell us which pathways are active, which substrates are available, which neighboring organisms will modify the products, which molecules will survive, or what chemistry will exist at a particular boundary and moment.

The species label hides the strain.
The strain genome hides the active state.
The ghost genome is the chemical reality that emerges when the living system runs.

Capacity is not state

Even the comparison *human genes versus microbial genes* preserves the idea that biology is understood once the code has been counted.

A metagenome can tell us that a community has the capacity to perform a reaction. A metatranscriptome can tell us that RNA associated with that capacity is present. A metaproteome can show that enzymes were made. None, by itself, guarantees the concentration, location, lifetime, transport, or biological effect of the resulting molecules.

Genes define possibilities. Biology selects among them.

The same microbial community can generate different chemical outputs when oxygen falls, pH shifts, a competitor arrives, a phage infects, the host becomes inflamed, or a dietary precursor appears. Enzymes can accept more than one substrate. Molecules can be modified by neighboring microbes, transformed by host tissues, or broken into active products. The final signaling entity may have no single author.

Suppose one bacterium carries genes that build an intermediate peptide. A second organism removes a residue. A host enzyme adds a sulfate. A dietary lipid supplies an acyl group. A local metal ion stabilizes a transient conformation. The active entity binds a receptor, is rapidly degraded, and never accumulates in blood at a concentration detectable by a conventional assay.

Where is that molecule encoded?

The honest answer may be: nowhere in one place.

It is encoded in the relationships.

This possibility is central to the ghost-genome hypothesis. The hidden chemical layer of life may not be a warehouse of stable compounds waiting to be cataloged. It may be a dynamic state space produced by interactions - a language whose words change with speaker, listener, location, substrate, and time.

The hardware bias in biology did not begin with negligence. It grew from success. The central dogma - DNA to RNA to protein - never claimed that metabolites were unimportant. Francis Crick's formulation concerned the direction of sequence information. But industrial drug discovery often converted a useful statement into a workflow: find the gene, identify the protein, build a molecule to bind it.

That workflow worked spectacularly in the right problems. It also encouraged the target to become the disease and the binding event to become the explanation. The rest of the organism became *complexity* to be managed later.

The chemical execution layer did not disappear. It simply became harder to fit into the diagram.

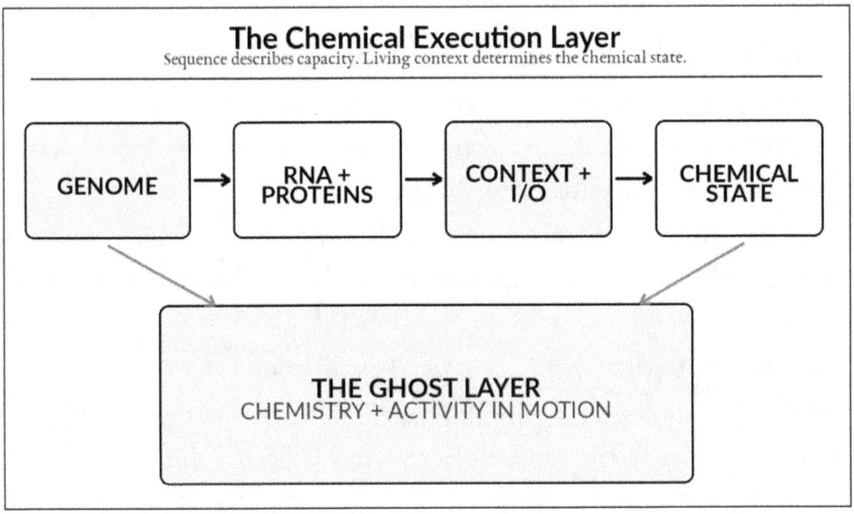

Figure 2. The chemical execution layer. Genes describe capacity; living context and input-output flows determine the real-time chemical state.

How to count a language that keeps changing

The practical problem begins with definitions.

A metabolomics instrument records features. A feature is commonly described by mass-to-charge ratio, retention time, intensity, isotope pattern, fragmentation, and sometimes ion mobility or spatial location. One molecule can generate several features through isotopes, adducts, charge states, fragments, dimers, or alternative conformations. Several molecules can also coelute into one apparent feature.

A formula is not a structure. A structure is not a stereochemical assignment. A stereochemical assignment is not a biological state.

A mature census of ghost chemistry would therefore need nested levels rather than one dramatic total.

Observed features are instrument-dependent signals that survive quality control and recur across samples or experiments.

Molecular families group related features through fragmentation, biosynthetic logic, exact mass differences, retention behavior, or other evidence.

Chemical entities distinguish covalent structures, constitutional isomers, stereoisomers, conjugates, and stable modifications.

Functional states include location, binding, metal coordination, redox state, complex formation, and conformations that alter biological activity even when covalent identity is unchanged.

Causal signals are the much smaller subset for which exposure, necessity, sufficiency, and mechanism can be established in a relevant system.

No honest program should collapse these layers.

The count also needs a sampling frame. One person's fasting blood at one moment is not the same object as the collective human-associated chemical repertoire across body sites, diets, geographies, medications, ages, diseases, and time. Stool favors what reaches the end of the intestine. Blood favors what survives absorption and clearance. Urine favors what the kidney excretes. Tissue homogenates favor stable material that remains after compartments are mixed.

Each sample is a biased witness.

The most important missing molecules may not be the rarest in absolute terms. They may be conditionally abundant - appearing after a meal, during inflammation, at a wound edge, within a biofilm, or during a phage event. Weekly sampling can make a repeated transient look nonexistent.

This is why accumulation curves matter. As a study adds people, time points, diets, tissue boundaries, and analytical methods, do new reproducible molecular families continue to appear after species and gene counts begin to saturate? If chemical discovery saturates quickly, the stronger ghost-genome claim weakens. If chemistry continues expanding under controlled perturbation, the case for context-generated output strengthens.

The two-billion hypothesis is useful only if it produces this kind of experiment.

Counting should reduce mystery, not brand it.

The business cost of counting the wrong layer

A scientific accounting choice eventually becomes a capital-allocation choice.

If disease is represented primarily as a defective target, companies build target-validation groups, screening libraries, medicinal-chemistry filters, and governance processes around that unit. If chemical function is represented primarily by known small molecules, assay collections and machine-learning systems inherit that boundary. If the microbiome is represented primarily by species abundance, diagnostic companies optimize sequencing and classification rather than functional chemistry.

Organizations become very efficient at producing the object their dashboard can recognize.

The hidden cost appears as attrition. A target can be genetically associated and still prove therapeutically intractable. A potent binder can fail to reach the relevant tissue. A disease category can contain multiple mechanistic subtypes. A molecule can look safe in an isolated system and become toxic after metabolism or community transformation. Each failure is explained locally, while the representation that generated the portfolio remains intact.

This is not an argument for abandoning targets, genes, or conventional libraries. It is an argument for treating them as layers in a larger evidence architecture.

A company that measures only the alphabet should not be surprised when it cannot predict the conversation.

The dashboard problem

The accounting error is not confined to science. It becomes a business problem whenever an organization mistakes what is easy to measure for what creates value.

A drug-discovery dashboard can count targets nominated, compounds screened, hits confirmed, structures solved, models trained, candidates generated, patents filed, or programs advanced. Every metric may be accurate. The portfolio can still become less productive if the counted activities are weakly connected to the difficult outcome: a medicine that changes a patient's life at an acceptable cost and risk.

The same distinction appears in artificial intelligence. A model can generate ten million structures and outperform another model on a benchmark. Those are real accomplishments. They do not establish that the generated set contains more viable medicines. If the benchmark rewards similarity to the historical record, the model may become increasingly skilled at reproducing the chemical preferences that created the record.

The trouble begins when proxy and purpose become indistinguishable.

Organizations need proxies because the final outcome arrives years later. A discovery team cannot wait for a Phase III trial before deciding whether this week's experiments were useful. The challenge is to choose intermediate measurements that preserve contact with the end state.

For the ghost-genome problem, a better dashboard would not count only peaks or predicted structures. It would ask:

- How many molecular features reproduce across independent collections?
- How many become structurally confirmed rather than tentatively annotated?
- How many carry source, spatial, temporal, and fraction metadata?
- How many biological activities survive orthogonal assays?
- How many candidate mechanisms include measured exposure in the relevant compartment?
- How often does a negative result change the next experiment?
- What is the cost per validated insight, not the cost per generated hypothesis?

Those questions are less glamorous than a billion-molecule virtual library. They are closer to the work that changes a base rate.

The lesson of GeneSweep is not that counting misleads. It is that a count becomes dangerous when the unit encodes the old theory. Before the Human Genome Project, the expected gene number carried an assumption about how complexity should be stored. In drug discovery, the number of candidates can carry an assumption that value is proportional to search volume. In microbiome science, a species count can carry an assumption that identity is an adequate proxy for function.

A strong dashboard exposes assumptions. A weak one hides them behind decimals.

The number that should not become dogma

The phrase *two billion ghost molecules* is designed to challenge the scale of our assumptions, not to pretend that a census has been completed.

No instrument has counted two billion unique functional molecules in the human microbiome. The number is a hypothesis about possible order of magnitude when several sources of diversity are considered together: microbial species and strain variation; enzyme promiscuity; host conjugation and deconjugation; dietary and environmental precursors; stereoisomers and oxidation states; community-dependent co-metabolism; transient intermediates; local reaction products; and changes across time, disease, medication, age, and geography.

The estimate could be too high. It could be too low. The answer depends on the counting rules. Should two stereoisomers count separately when only one is active? What about a molecule bound to a metal? A short-lived complex? A host-conjugated form? A metabolite that exists only in one micron of mucus for thirty seconds after a meal?

GeneSweep offers a warning here. The participants were not foolish because their estimates differed. They were confronting an object whose boundaries were still being defined. Birney's expanding footnotes were not bureaucratic clutter. They were evidence that the act of counting required a theory of what deserved to be counted.

The same discipline must apply to the ghost genome.

A mass-spectrometry feature is not automatically a molecule. A molecule is not automatically functional. A function is not automatically beneficial. A correlation is not causation. Any serious estimate must remove adducts, fragments, isotopes, contaminants, and analytical duplication while preserving meaningful structural and spatial diversity.

If the functional chemical layer is only modestly larger than current catalogs, improved annotation may close much of the gap. If it is orders of magnitude larger and heavily context-dependent, then the dominant discovery strategy - infer function from genes, screen isolated targets, and search synthetic space - is structurally incomplete.

We cannot choose between those futures by argument alone.

We need to look.

A Running System Is More Than Its Source Code

The distinction between capacity and state becomes easier to see if we borrow a careful analogy from computing.

DNA is the durable source code and low-level architecture. It defines components, interfaces, and capabilities that a cell can build. RNA and proteins are closer to loaded programs, services, and machines. They show which instructions are being read and which tools are available.

The chemical state is the running system. Metabolites, peptides, lipids, ions, conjugates, complexes, and molecular gradients are messages, inputs, outputs, control signals, memory states, and resource levels. They report what the system is doing now rather than only what it could do in principle.

Microbes, food, medicines, viruses, temperature, stress, and environment act like external inputs and co-processors. Tissue structure and past exposure provide system state. There is no conscious operator typing commands, but there is continuous feedback: outputs alter receptors and gene expression; those changes alter habitat and machinery; the changed system produces a different next output.

Mutation, recombination, horizontal gene transfer, developmental variation, and stochastic molecular events provide forms of exploration. Selection retains some changes and eliminates others. The result is not random in outcome, even though variation enters through processes that are not centrally planned.

A source-code listing can tell us what a program permits. It cannot tell us which user opened which file, what data arrived through an input channel, how memory changed, or what the system returned at a particular second. In the same way, a genome can reveal capacity without revealing the complete chemistry present at a mucosal boundary after one meal, one medicine, and one microbial disturbance.

Bile acids show the full stack. Human genes build primary bile acids. The liver conjugates and transports them. Microbial enzymes edit them. Diet and anatomy shape the route. Host receptors read the edited products and change future metabolism. The meaningful unit is not one gene or one molecule. It is the executed relay.

The same is true for tryptophan, polyphenols, drugs, and cell-derived material. The chemical novelty may reside in a new structure. It may also reside in the sequence of authors, the location, the timing, the conformation, or the combination.

Genomics remains indispensable. The correction is structural: sequence describes the possibility space; chemistry reports the execution. The ghost genome lives in the difference.

Why the accounting error changes medicine

An incomplete inventory would be a technical inconvenience if the missing chemistry had little biological consequence. The known examples make that position difficult to defend.

Microbial metabolites can activate human G-protein-coupled receptors. They can influence nuclear receptors, immune differentiation, epithelial barrier function, vascular biology, and drug response. Short-chain fatty acids can be fuel, receptor ligands, and epigenetic modifiers. Bile-acid derivatives can shape both microbial competition and host signaling. Aromatic metabolites can be protective in one context and harmful when clearance fails. Cell-wall fragments can train immunity, trigger inflammation, or do both at different doses and times.

The point is not that every process is controlled by microbes. Human biology remains human biology. The point is that the operating system is hybrid. When medicine treats microbial chemistry as background noise around a human target, it may remove part of the mechanism from the experiment before asking what the mechanism is.

This is especially consequential in chronic disease. An acute infection can sometimes be traced to a pathogen and treated by removing it. A chronic inflammatory, metabolic, neurodegenerative, or fibrotic state is often sustained by feedback across tissues, immunity, behavior, circulation, microbes, and environment. A single target may still be therapeutically powerful, but success at one target does not prove that the disease is one target.

The accounting error also affects safety. A compound optimized against a purified protein may behave differently in a body where microbial enzymes transform it, transporters redirect it, and local chemical states change its exposure. Conversely, a molecule dismissed as too large, too polar, or too multi-targeted under conventional rules may operate effectively when its action is local, transient, or conformationally adaptive.

The missing chemistry therefore sits on both sides of the drug-development equation. It may contain new therapeutic starting points, and it may explain why apparently rational candidates fail.

GeneSweep's most important lesson is not that experts once overestimated a number. It is that a successful measurement can force a field to revise the relationship between its parts and its whole. The human genome did not shrink human complexity. It made context impossible to ignore.

The ghost genome asks whether we are approaching the same moment in chemistry.

What the wager teaches AI

The betting book returns later in this story because it is a nearly perfect parable for artificial intelligence.

A model can contain a vast amount of correct information and still be trained around the wrong unit of importance. The Human Genome Project did not produce bad data. It produced extraordinary data. The conceptual error came from expecting a count of durable instructions to explain the full behavior of a dynamic organism.

AI systems trained on modern chemical databases face a related problem. Their records contain enormous knowledge, but the records are shaped by what previous generations chose to synthesize, assay, publish, patent, and retain. If context-dependent natural products, large macrocycles, unstable metabolites, negative results, and transient molecular states are underrepresented, the model will learn that absence as though it were reality.

The first task is therefore not to make the model imagine more.

It is to make the world measurable in a form the model can learn.

That is the lesson hidden in GeneSweep's plastic cup. Before we can count the chemical language of life, we have to decide whether we are counting the alphabet, the words, the sentences, or the conversation.

The next story begins with a pathologist looking through a microscope at a spiral shape in a place medicine had declared too hostile for bacteria.

The observation was small.

The disease it would rewrite was not.

CHAPTER 2

The Doctor Who Drank the Evidence

When a Bacterium Rewrote a Disease

STORY 2

THE DOCTOR WHO DRANK THE EVIDENCE

When a Bacterium Rewrote a Disease

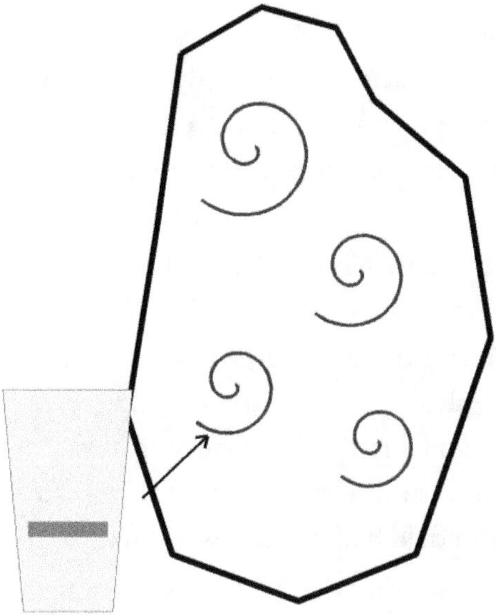

THE TURNING POINT

A chronic disease changed category because an observation survived testing.

Robin Warren was not looking for a revolution. He was looking through stomach-biopsy slides.

In the late 1970s, the Australian pathologist began noticing curved and spiral-shaped bacteria clustered near inflamed stomach tissue. The observation was awkward. The stomach was understood as an acid bath, hostile to sustained bacterial life. When organisms appeared in specimens, contamination or irrelevance was a more comfortable explanation than residence.

Warren kept looking.

The bacteria were not scattered randomly across every slide. They seemed associated with inflamed mucosa. Pattern is where many scientific stories begin: not proof, not mechanism, but a repetition that refuses to behave like noise.

In 1981, a young physician named Barry Marshall needed a research project during his training at Royal Perth Hospital. Warren needed a collaborator willing to chase the spiral organism. Marshall later described himself as an unlikely medical revolutionary. He had grown up reading his father's technical books and his mother's nursing texts, moved through several medical specialties, and retained the generalist's willingness to become interested before a field had decided what mattered.

His original ambition was modest. Perhaps the organism was new. Perhaps he and Warren could describe it in a paper.

The disease model around them was anything but modest.

Peptic-ulcer disease had an established story. Stress, lifestyle, and excess acid damaged the stomach or duodenum. Acid-suppressing medicines could provide dramatic relief. Surgeons could cut nerves or remove portions of the stomach in severe cases. The model was biologically plausible, clinically useful, and commercially coherent.

It was also incomplete.

Marshall and Warren examined biopsy samples from patients undergoing endoscopy. The association between the curved bacteria, gastritis, and ulcers strengthened. Yet association was not causation. The organism had to be cultured, characterized, and linked to disease in a way the field could trust.

The culture plates repeatedly appeared negative.

The laboratory followed the usual incubation timetable for organisms suspected at the time. After a few days without colonies, plates were discarded. Then an Easter holiday interrupted the routine. Some plates remained in the incubator longer than planned. When the team returned, colonies had appeared.

The bacteria had not been absent. The experiment had been ending too early.

This detail matters far beyond the history of *Helicobacter pylori*. A standard method does not merely reveal nature. It defines the window during which nature is allowed to appear. Change the oxygen, nutrients, incubation time, sample size, extraction solvent, ionization mode, or threshold, and the visible world changes with it.

The culture gave Marshall and Warren stronger evidence, but it did not instantly make the disease story legible to everyone else. They encountered skepticism, difficulty publishing, and the familiar demand for more convincing causal proof. The skepticism was not irrational. The history of medicine is full of organisms found near disease that did not cause the disease. The stomach is exposed to swallowed microbes constantly. A new bacterium, an inflammatory lesion, and a correlation were not enough.

The problem was how to obtain the next level of proof before the field accepted the premise required to fund it.

Marshall had treated some patients with antibiotics and bismuth and seen improvement, but he lacked the kind of controlled evidence that could overturn an entrenched model. He eventually chose an experiment no institutional review board would encourage today: he swallowed a culture of the organism himself.

Within days he developed nausea, bad breath, vomiting, and biopsy-confirmed gastritis. The bacterium was recovered from his stomach. Antibiotics cleared the infection.

The self-experiment did not, by itself, prove that *H. pylori* caused every ulcer. Marshall developed gastritis, not a fully established peptic ulcer during the brief experiment. The decisive case accumulated through pathology, culture, epidemiology, treatment studies, recurrence data, and clinical trials. But the act forced the hypothesis into a human body in a form that was difficult to dismiss as contamination.

The old disease story began to change.

Many peptic ulcers could be treated not by lifelong suppression of acid alone but by eradicating an infection. The discovery eventually brought Marshall and Warren the 2005 Nobel Prize in Physiology or Medicine. More importantly, it converted a chronic, recurring condition in millions of patients into a disease with a curative strategy in many cases.

The bacterium had been in the stomach all along.

The missing element was permission to treat the observation as causally important.

Why the old explanation was so durable

The acid-and-stress model survived because it contained important truths.

Acid can damage tissue. Reducing acid can relieve pain and allow ulcers to heal. Smoking, severe physiological stress, and medication can influence risk. Surgical and pharmaceutical treatments produced observable benefit. The model did not persist merely because physicians were incurious. It persisted because patients improved when the model was used.

That is what makes incomplete models difficult to displace.

A treatment can work downstream of the initiating cause. A dam can lower the water level without explaining why the river is flooding. Acid suppression reduced the chemical injury experienced by an ulcer. It did not necessarily remove the inflammatory process that made recurrence likely.

By the early 1980s, H2-receptor antagonists such as cimetidine had become major medicines. They represented genuine progress over surgery and uncontrolled symptoms. A microbial explanation threatened neither the fact that these drugs worked nor the biological importance of acid. It threatened the idea that chronic suppression was the natural endpoint of treatment.

The economic architecture followed the clinical architecture. Recurring disease supported recurring medication. A short course of antibiotics capable of eliminating an upstream cause changed the product logic as well as the medical logic.

This is where institutional resistance can become morally simplified after the fact. It is easy to imagine a drug industry protecting revenue or a medical establishment protecting authority. Those incentives can exist. They are not required to explain the delay. A field with effective symptom control, a plausible mechanism, and decades of training will demand unusually strong evidence before accepting that its causal hierarchy is wrong.

The responsible question is not why everyone failed to believe immediately.

It is why the system had so few efficient ways to test an anomaly that could change the entire treatment model.

The chain of evidence

Marshall and Warren's work became convincing because the evidence accumulated across methods.

Pathology showed the organism near inflamed mucosa. Clinical sampling showed strong associations with gastritis and ulcer disease. Culture made the organism available for study. Treatment and recurrence patterns showed that removing the bacterium changed the disease course. Later work clarified virulence factors, host response, transmission, cancer risk, and the varied outcomes of colonization.

Each layer ruled out an alternative explanation.

The bacterium was not merely present because an ulcer existed. It could precede disease and drive inflammation.

The improvement was not merely an antibiotic's nonspecific anti-inflammatory effect. Eradication status predicted recurrence.

The association was not confined to one unusual hospital population. It appeared across countries and cohorts.

The organism was not an all-purpose villain. Many infected people remained asymptomatic, and the relationship between strain, host, environment, and disease varied.

This is the causal ladder the ghost-genome field will need.

Association begins the investigation. Structural identity defines the suspect. Plausible exposure shows the molecule can reach the relevant site.

Adding the molecule tests sufficiency. Removing the molecule, enzyme, or producer tests necessity. Rescue strengthens the chain. Close analogs and orthogonal assays test specificity. Replication outside the originating laboratory makes the claim portable.

The ladder is demanding because complex systems generate correlations effortlessly.

A molecule can rise after tissue damage without causing it. A microbial species can expand because medication changed the habitat. A metabolite can mark a diet while contributing nothing to the outcome. A feature can track disease because kidney clearance failed.

The more compelling the story, the more important the ladder becomes.

The experiment in the mirror

By 1984, Marshall believed the evidence had reached an infuriating stage: strong enough to disturb the accepted ulcer story, not yet strong enough to reorganize practice.

The bacterium had been seen in tissue. It had been cultured. It was associated with gastritis. But association could still be dismissed. Perhaps the organism colonized an already damaged stomach. Perhaps it was an incidental passenger. Perhaps the staining, culture, or pathology was being overinterpreted by investigators who had become attached to their own anomaly.

Marshall decided to make the causal chain more direct.

He underwent endoscopy to document that his stomach was not already inflamed, then swallowed a culture of the organism. The act has acquired the cinematic simplicity of scientific folklore: a doctor drinks bacteria and proves the establishment wrong. The real sequence was more instructive.

Several days later, Marshall became unwell. His breath changed. He developed nausea and vomiting. A repeat endoscopy showed acute gastritis, and the organism was recovered. He treated himself with antibiotics.

The experiment did not reproduce the full natural history of a chronic ulcer. It did not satisfy every modern ethical expectation. It did not prove that the bacterium explained every case. What it did was collapse one of the easiest defenses: the organism could no longer be treated only as a harmless inhabitant of damaged tissue.

The drama of the self-experiment helped the story travel. The science traveled because other people could test it.

Patients with ulcer disease received antimicrobial regimens. The organism disappeared in many of them. Relapse rates changed. Pathology improved. The observation moved through the causal ladder: association, intervention, response, replication.

This is the difference between a compelling anecdote and a scientific reversal. The anecdote creates attention. The reversal survives outside the person who told it.

The story also reveals a recurring burden placed on early discoveries. The new claim is often asked to provide a complete replacement model immediately, while the old model is allowed to remain useful despite known gaps.

Acid suppression clearly helped ulcers. That fact made the acid-centered model clinically coherent. A microbial explanation had to account not only for the bacterium but for the role of acid, tissue injury, host response, and recurrence. The new theory had to be more complete than the old theory had ever been required to be.

This asymmetry appears whenever hidden chemistry challenges a target-centered model. A conventional molecule can advance because it binds a known target and produces a familiar assay result. A context-dependent microbial molecule may be asked to identify its producer, exact structure, spatial concentration, target pattern, causal mechanism, patient subgroup, manufacturing route, and clinical relevance before the platform needed to answer those questions is funded.

Those questions are legitimate. The timing can be circular.

Marshall's self-experiment was a drastic way of breaking the circle. The better modern answer is to design discovery systems in which unusual observations can earn the next experiment without being mistaken for established truth.

A chronic disease can have a microbial cause without becoming simple

H. pylori also teaches restraint in the opposite direction.

Once microbes became scientifically fashionable, nearly every chronic disorder acquired a proposed microbial connection. Association studies found altered communities in obesity, autism, depression, cancer, autoimmune disease, and neurodegeneration. Some signals were reproducible. Many were confounded by diet, medication, geography, age, and disease severity. The microbiome became an answer before it had become a causal mechanism.

The proper lesson from H. pylori is not that every chronic disease has one hidden pathogen. It is that the boundary between infectious, inflammatory, metabolic, and environmental disease can be more porous than our specialties imply.

A microbe can cause disease directly. A community can alter risk. A metabolite can change treatment response. An infection can trigger an immune state that persists after the organism is gone. Microbial absence can matter as much as presence. The same organism can be harmless in one host and pathogenic in another.

The causal unit may be an organism, a gene, a molecule, a community state, or a sequence of events.

Medicine needs the freedom to discover which one without assuming the answer in advance.

The system that funds yesterday

When I began describing a platform intended to recover chemistry that ordinary sampling and culture methods could not reliably reach, the question arrived in one form after another:

What is the clinical application?

It was a reasonable question. Investors, grant reviewers, and pharmaceutical teams cannot fund an abstraction forever. A clinical application defines a patient, a market, an experiment, and a development path.

But we were being asked to name the destination before the telescope had been built.

The proposed platform was not a drug candidate. It was a way to reach chemistry that conventional workflows might never make experimentally available. If the premise was correct, the first important result might emerge from the library rather than precede it.

The circularity was hard to explain in a room organized around assets. To fund the instrument, we were expected to show the asset. To show the asset, we needed the instrument.

There is no secret room where a committee votes to stop the future. That would make the story easier. It would give us villains with names, minutes, and motives. The real mechanism is more durable because it requires no conspiracy. It is built from reasonable people making defensible decisions inside a system that punishes the wrong kind of error.

Consider a grant reviewer faced with two proposals.

The first extends a familiar model. It uses accepted assays, a well-studied target, an experienced team, and a clear statistical plan. The effect may be incremental, but the path is legible.

The second proposes that the accepted model is incomplete. It crosses fields. The decisive assay does not yet exist. The preliminary data are suggestive but strange. Success could open an entirely new direction; failure could produce a pile of uninterpretable measurements.

Which proposal is easier to score?

Peer review was created to protect science from whim, fraud, and charisma. It asks knowledgeable people to evaluate whether a question matters, whether the methods are rigorous, and whether the team can execute. No serious research system can abandon those responsibilities.

But a method that is excellent at judging work within a paradigm may be poorly calibrated to judge work that changes the paradigm's boundaries.

The problem is not that reviewers hate novelty. Evidence on novelty and funding is more complicated. Some novel work is rewarded. High-risk programs exist precisely because ordinary review can struggle with early, cross-disciplinary, or difficult-to-compare ideas. The evidence does not support a cartoon in which every reviewer reflexively rejects every new idea.

It does support a structural tension.

A panel must make decisions using the information available now. Transformative work often becomes obvious only after the missing instrument, dataset, or conceptual bridge has been built. Reviewers are asked to predict that future while being held accountable for present-day rigor. The safe response is to reward feasibility.

Feasibility is not the enemy. But when feasibility becomes the dominant currency, the system funds the next experiment more easily than the next way of seeing.

The preliminary-data loop

Innovators are frequently told to return with preliminary data. The request sounds neutral. In practice it can create a circular gate.

To obtain the data, the investigator needs money. To obtain the money, the investigator needs the data.

Established laboratories can bridge the gap with discretionary funds, equipment, philanthropy, networks, and the credibility of prior success. Outsiders and small teams often cannot. The result is an invisible tax on ideas that require a new platform before they can produce evidence in the format expected by the old platform.

Microbiome chemistry intensifies the problem. Imagine a researcher who believes a low-abundance molecule produced only in an intact community influences neuroinflammation. A conventional panel may ask for the structure, concentration, target, mechanism, exposure, reproducibility, and disease relevance. Those are legitimate scientific questions.

Answering them may require collecting an ecosystem at unusual scale, preserving context, separating thousands of compounds, acquiring high-resolution spectra, building analytical tools, synthesizing candidate structures, and testing them across host and microbial models. The proposal is not one experiment. It is an infrastructure bet.

The old system hears: too many unknowns.

The innovator hears: that is the point.

Marshall and Warren confronted a smaller version of the same loop. The evidence required to justify the new disease model depended on treating the bacterium as more than contamination. But the resources and attention required to generate that evidence were constrained by the prevailing assumption that the stomach was not a microbial habitat and that acid was the principal story.

The Easter culture plates broke the loop because time, almost by accident, became a new instrument.

Experts, scar tissue, and the map

The strongest critique of institutions is often expressed as accusation: reviewers protect their careers, incumbents defend their products, and specialists suppress ideas that threaten prestige. Human beings are capable of all of that. The more useful explanation is subtler.

Expertise is a prediction engine built from experience. A scientist who has spent thirty years studying a pathway has seen hundreds of seductive claims collapse. Skepticism is not only self-interest; it is scar tissue. The expert knows which controls are usually missing, which measurements drift, which animal models mislead, and which revolutions quietly disappear after the press release.

That knowledge is indispensable.

It also creates a blind spot. The expert's filter is optimized for false positives inside the known system. A genuinely new system may look exactly like the false positives the expert learned to reject.

This is the institutional version of the innovator's dilemma. Good organizations fail not because they stop listening but because they listen carefully to the signals their current architecture was designed to value.

A pharmaceutical company values a target with a validated assay, a molecule with a known mechanism, an intellectual-property path, and a regulatory precedent. A university values publications, grant continuity, and questions that can be completed within a trainee's funding cycle. A public agency values accountability and defensible selection.

None of those values is irrational. Together they can create a machine that improves yesterday with exceptional discipline.

The challenge is not to replace expertise with intuition. It is to separate two questions that institutions often collapse:

1. Is the present evidence sufficient to believe this claim?
2. Is the observation important enough to justify building the experiment that could make the evidence sufficient?

Marshall's self-experiment is memorable because he answered the second question with his own stomach. That should not be the model for responsible modern research. The institutional lesson is that no scientist should need to become the assay simply because the system cannot fund the bridge between anomaly and proof.

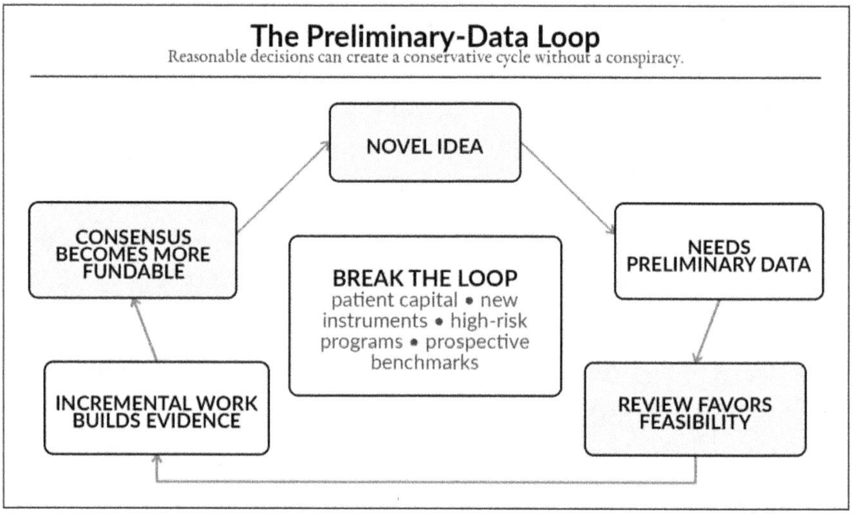

Figure 3. The preliminary-data loop: a rational incentive cycle can become conservative without anyone intending it.

A field can be right and still become narrow

The H. pylori story is often used as a weapon against mainstream science: the experts were wrong, the outsider was right, therefore today's dissident deserves belief. That conclusion is not only false; it destroys the discipline that made the original reversal possible.

The more useful lesson is that a field can be directionally right and strategically narrow at the same time.

Alzheimer's disease offers a modern example. For decades, amyloid-beta biology dominated attention and capital. The model was not invented from nothing. Genetics implicates amyloid processing. Plaques are a defining pathological feature. A vast body of work revealed real mechanisms. More recently, antibodies such as lecanemab and donanemab demonstrated that removing amyloid in carefully selected patients with early disease can modestly slow clinical decline.

Those results matter. They show that amyloid is not irrelevant.

They also show the limits of a monoculture. The treatments are not cures. They require early diagnosis, infusion infrastructure, imaging, and management of risks that can include edema and bleeding. The disease still involves tau, vascular injury, metabolism, immune activity, synaptic loss, aging, and likely multiple interacting routes to degeneration.

The lesson is not *amyloid was wrong*. It is that a correct component can become an overly complete story.

The same pattern appears elsewhere. Oncology can focus on a driver mutation and underestimate the tumor ecosystem. Autoimmune research can identify a cytokine and suppress it without fully understanding the loss of tolerance upstream. Metabolic medicine can target a hormone pathway while the patient remains a moving network of food, microbes, liver, muscle, adipose tissue, brain, sleep, behavior, and social environment.

Single targets can produce excellent drugs. The mistake is treating the success of a target as proof that the disease is a target.

H. pylori changed medicine because it forced a causal reordering. Acid still mattered, but it was no longer always the initiating explanation. A bacterium could sit upstream of the inflammation and recurrence that acid suppression managed downstream. In other diseases, the missing upstream factor may not be one organism. It may be a chemical state produced by a community.

Safe failure

Institutions do not merely select science. They select career risk.

A program manager who funds a conventional project that fails can say the hypothesis was well supported and the experiment was properly executed. A manager who funds a strange platform that fails may be asked why public money supported speculation.

The asymmetry is powerful: conventional failure is unfortunate; unconventional failure can become evidence of poor judgment.

This produces what might be called *safe failure.* Organizations repeatedly choose projects that are defensible even when portfolio-level results are disappointing. The same behavior appears in corporate research, venture investment, and government programs. Each individual decision can be rational while the aggregate trajectory remains Eroom's Law.

The corrective is not to replace peer review with intuition, celebrity, or random funding. It is to design portfolios that recognize different kinds of uncertainty.

Some resources should optimize known mechanisms. Some should test sharply defined hypotheses. Some should build missing instruments. Some should fund people rather than projects. Some should reserve capital for claims that would be transformative if true and inexpensive to falsify if false. Some should use reviewers from outside the threatened specialty. Some could use partial lotteries among proposals that clear a quality threshold, acknowledging the false precision of rankings among uncertain ideas.

Science needs both courts and frontiers.

The court asks: Is the evidence sufficient?

The frontier asks: What evidence would become possible if someone were given permission to build the telescope?

Systems can be redesigned. The NIH Pioneer, New Innovator, Transformative Research, and Early Independence programs are acknowledgments that unconventional work can struggle under ordinary review. Investigator-centered philanthropy, ARPA-style programs, venture philanthropy, and ring-fenced corporate groups all attempt, in different ways, to protect a question before the ordinary asset gates can evaluate it.

The first reform is intellectual honesty. We should stop pretending that the top-ranked incremental proposal and the top-ranked paradigm challenge can be evaluated with the same ruler.

The second is measurement. Funding systems should track not only publications and citations but how often portfolios create new tools, open new fields, falsify costly assumptions, or produce data that change the allocation of capital.

The third is cultural. Experts should be asked not only, *Is this proposal correct?* but also, *If it were correct, what part of our map would have to change?* The people who know the old system best are often the people most capable of identifying where a new claim genuinely conflicts with it.

The final reform is permission - not permission to believe without evidence, but permission to build the experiment that evidence requires.

The clinical revolution looked ordinary

The most important consequence of the discovery was not the Nobel ceremony or the famous glass of bacteria. It was a change in what happened to an ordinary patient.

Before the causal role of *H. pylori* was accepted, many people with peptic-ulcer disease entered a cycle of symptoms, acid suppression, dietary caution, recurrence, and sometimes surgery. The treatment model managed an enduring condition.

Once the infection could be diagnosed and eradicated, many of those patients entered a different story. A defined course of several medicines could remove a major cause of the disease and sharply reduce recurrence. The revolutionary idea arrived in the clinic as prescriptions, follow-up testing, and fewer repeat ulcers.

That is a useful preview of what a ghost-genome medicine would have to become.

The discovery path may involve an unusual ecosystem, an unknown molecular family, a network phenotype, and a new measurement architecture. The product cannot remain mysterious. It must eventually enter the familiar disciplines of identity, dosing, manufacturing, toxicology, diagnosis, clinical trial design, and patient selection.

The origin can be revolutionary. The delivery has to become dependable.

The *H. pylori* story also warns against a second simplification. Once a microbial cause was established, it became tempting to retell the old era as irrational and the new era as complete. Neither is accurate. Acid matters. Nonsteroidal anti-inflammatory drugs matter. Bacterial strain, host susceptibility, smoking, and other factors matter. Infection creates cancer risk in some settings, while many infected people never develop severe disease.

A better map does not replace complexity with a new slogan. It connects more of the evidence.

What *H. pylori* did - and did not - prove

The cleanest version of the story is also the least accurate: doctors thought stress caused ulcers; two rebels discovered bacteria; antibiotics cured everything.

Acid still matters. Nonsteroidal anti-inflammatory drugs cause ulcers. Not every infected person develops disease. *H. pylori* strains differ, hosts differ, and cancer risk, inflammation, and clinical outcomes depend on context. Acid-suppressing medicines remain useful. The microbiome did not abolish physiology.

The discovery changed the causal hierarchy. A microbe could live in the stomach, produce chronic inflammation, and sit near the center of a disease medicine had framed primarily through host chemistry and behavior.

That change blurred a boundary that still shapes biomedicine: the boundary between infectious and chronic disease.

Once a bacterium could help drive a long-term inflammatory disorder, other questions became more legitimate. Could microbial activity participate in cancer, vascular disease, autoimmunity, metabolic illness, or neurodegeneration? In some cases, evidence would remain weak. In others, microbes would prove causal, contributory, protective, or treatment-modifying. The important change was not that microbes became the answer to every disease. It was that medicine could no longer assume they belonged outside the chronic-disease model.

For the ghost genome, the lesson goes one step further. A bacterium is easier to identify than the chemical conversation it conducts with its host. The organism can be cultured, sequenced, imaged, and eradicated. The molecules through which it changes acid, inflammation, epithelial behavior, and neighboring organisms may be numerous, transient, and distributed.

The next frontier is not merely asking which microbe is present.

It is asking what chemistry the relationship produces.

Belief Disciplined by Falsification

The romance of the outsider can become dangerous. Opposition does not prove correctness, and conviction is not evidence.

Yet discovery often requires belief before proof is complete. Instruments, cohorts, collections, and new assays can take years to build. No one commits that effort while believing the question is merely decorative.

The responsible form of belief is a commitment to decisive tests. It begins by naming what the prevailing model gets right. It separates observation from interpretation. It states the strongest alternative explanation. It defines what result would reduce confidence. It invites experts to attack the experiment rather than defend the institution.

A serious frontier program should be able to say: Here is the trace. Here is how often it repeats. Here is the context in which it appears. Here is the control that should make it disappear. Here is the independent laboratory that can try to reproduce it. Here is the point at which we will stop.

This is not weaker science. It is stricter science applied earlier, when the instrument and the category are still being invented.

Marshall and Warren were not vindicated because they were stubborn. They were vindicated because the bacterium, the pathology, the response to treatment, and the recurrence pattern survived challenge. Their belief kept the investigation alive long enough for the evidence to become portable.

The ghost-genome thesis deserves the same bargain. Vision can justify building the observatory. Only reproducible chemistry and causal biology can justify the conclusion.

Reality did not care which ulcer explanation was more respectable. The stomach contained a microbial world medicine had not learned to see.

The next story leaves the body entirely. It begins with a star fifty-one light-years away, wobbling so slightly that almost no one could see it - and with a planet orbiting where the prevailing theory said a planet like it should not be.

CHAPTER 3

The Planet That Should Not Exist

A Modern Galileo Story

STORY 3

THE PLANET THAT SHOULD NOT EXIST

A Modern Galileo Story

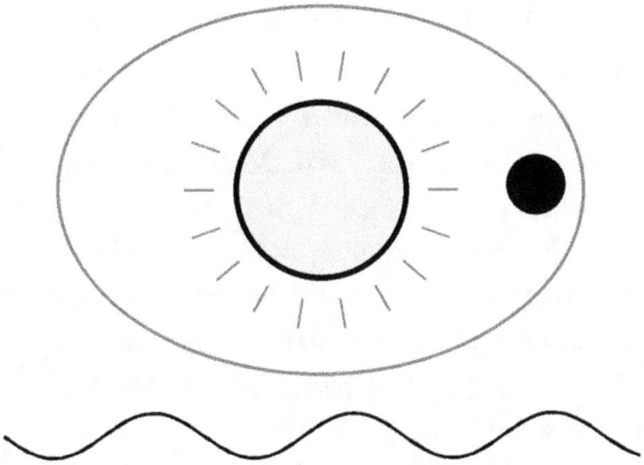

THE WOBBLE WAS THE EVIDENCE

THE TURNING POINT

The wobble forced the map to make room for a new kind of world.

The planet was invisible.

What Michel Mayor and Didier Queloz could see was a star moving toward them and away again with exquisite regularity.

The instrument was ELODIE, a new high-resolution spectrograph installed at the Haute-Provence Observatory in southern France. Its purpose was to measure the radial velocity of stars - the slight shift in their light as they moved toward or away from Earth. A planet does not orbit a perfectly fixed star. Planet and star circle a shared center of mass. The planet makes the larger journey; the star performs a small answering wobble.

For a Jupiter-like planet at a Jupiter-like distance, the wobble requires patience. One orbit takes nearly twelve years. Mayor and Queloz were looking for repeated shifts in starlight precise enough to betray such companions.

The signal around the star 51 Pegasi repeated every 4.23 days.

That was the problem.

The data suggested a planet with roughly half Jupiter's mass orbiting astonishingly close to its star - far inside Mercury's orbit in our own solar system. The object would be heated to temperatures no familiar giant planet experienced. Under prevailing models, gas giants formed in cold regions far from their stars, where enough material could accumulate. A Jupiter pressed against a Sun-like star was not merely an unexpected example. It threatened the architecture by which astronomers believed planetary systems assembled.

When an observation makes no theoretical sense, the first suspect is the observation.

Could the star be pulsating? Could rotating starspots mimic the velocity signal? Could the instrument be drifting? Had a calculation gone wrong? Queloz, then a doctoral student, had to distinguish a planet from every more ordinary way a star or spectrograph could fool them.

The signal remained.

On October 6, 1995, Mayor and Queloz announced the first confirmed planet orbiting a Sun-like star. The planet, now known as 51 Pegasi b, became the prototype of a class that had barely existed in theory: the hot Jupiter.

The discovery did not simply add one planet to a catalog. It forced planetary science to explain migration, orbital evolution, and the startling diversity of systems that followed. Once instruments improved and surveys expanded, hot Jupiters became only one category among super-Earths, mini-Neptunes, resonant systems, eccentric giants, and arrangements that made our solar system look less like a template and more like one possibility.

The planet had always been there.

The model had no place for it.

The wobble had to survive every ordinary explanation

Radial-velocity discovery is an exercise in disciplined suspicion.

A spectrograph separates starlight into a spectrum crossed by absorption lines. When the star moves toward Earth, the lines shift slightly toward shorter wavelengths; when it moves away, they shift toward longer wavelengths. The motion caused by a planet is tiny compared with the speed of light and can be entangled with instrument drift, stellar rotation, convection, pulsation, and spots on the star's surface.

ELODIE was built for repeatability. Calibration and a wide spectral range allowed Mayor and Queloz to compare many lines and look for a periodic signal. The four-day rhythm around 51 Pegasi was too fast for the Jupiter analog they expected and too coherent to ignore.

A planet was not the only possible interpretation. Stellar pulsation could create apparent motion. A rotating surface feature could distort spectral lines. The team looked for changes in line shape and other indicators that would distinguish a companion from an active star. Independent groups moved quickly to confirm the signal.

The result changed theory because it was not absorbed as a one-off curiosity. If a gas giant could orbit so close to a Sun-like star, then either giant planets could form in conditions theorists had not considered or they could move after formation. Planetary migration became essential. The architecture of a solar system was no longer assumed to be a fixed consequence of where its planets first assembled.

This is how an anomaly becomes a field. It survives artifact checks, appears elsewhere, and demands a mechanism that explains both the old observations and the new one.

The ghost genome will require the same progression. One unusual spectrum is an invitation. Recurrent molecular families across samples are stronger. A structure confirmed by synthesis is stronger still. A causal phenotype, exposure, and mechanism transform the anomaly into biology.

A four-day year

The number that made 51 Pegasi b feel impossible was not its mass. It was its calendar.

The planet completed an orbit in just over four days. A gas giant, something in the broad family of Jupiter, was circling so close to its star that an entire year passed between one Monday and the end of the workweek.

At Haute-Provence Observatory, the ELODIE spectrograph did not photograph the planet. It measured changes in the star's spectrum as the star moved slightly toward and away from Earth. The signal was small, periodic, and vulnerable to ordinary explanations. Stellar activity can imitate motion. Instruments drift. Sampling patterns create aliases. A claim this consequential had to survive the suspicion that the wobble was in the equipment or the star rather than caused by an unseen companion.

Mayor and Queloz kept measuring.

The repeated 4.23-day rhythm became harder to dismiss. The inferred companion was too massive and too close to resemble the architecture on which many planetary models had been built. In the Solar System, the giant planets occupy the cold outer regions. The prevailing formation story placed them there for a reason: large cores and gaseous envelopes were expected to assemble beyond the warmer inner disk.

The new planet was not merely another object. It was a configuration the accepted map did not comfortably produce.

The eventual explanation did not require abandoning physics. It required adding history. A giant planet could form farther out and migrate inward through interactions with the disk or other bodies. The observation expanded planetary dynamics.

That pattern matters for biology. When a molecule appears in a place, form, or concentration the accepted pathway does not predict, the first temptation is often to protect the map by treating the signal as contamination, artifact, or exception. Sometimes that is exactly correct. Instruments produce artifacts. Samples are contaminated. Biology is noisy.

But a repeated anomaly can contain a missing history.

A microbial metabolite in blood may have been transformed by the liver. A peptide at a wound edge may have been cleaved from a larger host protein and edited by microbial enzymes. A molecule absent in pure culture may require a neighbor's substrate. A spectral family that looks biosynthetically impossible in one organism may be the result of a relay.

The equivalent of planetary migration is context: a process that moved the product from the place our simple model expected into the place the instrument found it.

The right question is not, *Does this observation violate the model?*

It is, *What history would make both the observation and the model's valid pieces true?*

The Galileo Effect

The popular story of Galileo is polished by centuries of retelling. Learned men are said to have refused the telescope. Galileo is said to have whispered *And yet it moves* after recanting. Historians debate the details.

The story endures because it captures a recurring truth: a new instrument can make old certainty expensive.

By the time Galileo turned an improved telescope toward the sky in 1610, debates about the heavens had lasted for centuries. The telescope changed the argument because it introduced observations the inherited model did not comfortably predict - mountains on the Moon, phases of Venus, and moons orbiting Jupiter.

The telescope did not make Galileo courageous. It made denial harder.

I call the recurring pattern the Galileo Effect. It usually unfolds in four stages.

First, a new instrument appears. It can be physical, like a telescope or microscope; conceptual, like natural selection; or computational, like sequence comparison.

Second, the instrument reveals an anomaly that the prevailing model treats as noise, artifact, exception, or irrelevance.

Third, the observer is asked to translate the anomaly into the old language before the new language has been accepted.

Fourth, after enough evidence accumulates, the anomaly becomes part of a revised field - and the revision is remembered as though it had been inevitable.

51 Pegasi b is a modern Galileo story because the discovery depended on precision, not spectacle. No one looked through an eyepiece and saw a giant world glowing beside its star. The evidence was a periodic pattern in numbers. The observer had to trust the instrument enough to investigate the anomaly and distrust it enough to eliminate every artifact.

That tension is the proper posture for the ghost genome.

An unknown mass spectrum is not a revelation merely because it is unknown. A molecular feature can be an adduct, isotope, contaminant, in-source fragment, or processing artifact. A microbial association can reflect diet, medication, geography, or disease rather than causation. A computational structure can be confidently wrong.

The instrument earns authority through calibration, controls, repetition, and independent confirmation.

But once those tests are passed, the model must be allowed to change.

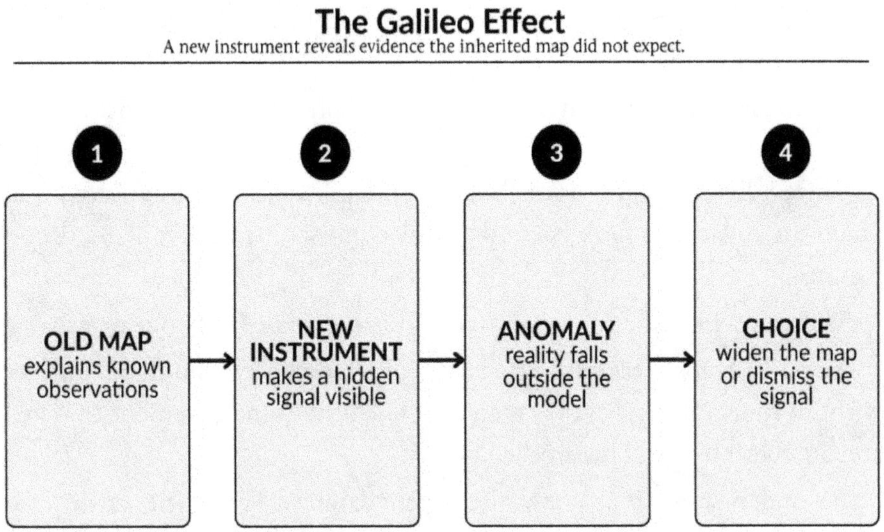

Figure 4. The Galileo Effect: old map, new instrument, anomaly, and the choice to widen the model or explain away the signal.

Carl Woese changes the unit of life

Microbiology had its own version of the invisible planet.

For much of the twentieth century, classification relied heavily on shape, metabolism, staining, and the organisms scientists could culture. Bacteria were diverse, but they occupied one broad category of simple prokaryotic life. The map looked stable because the instruments reinforced it.

Carl Woese chose a different measurement. He compared ribosomal RNA, a molecular component shared across cellular life whose sequence changes could preserve deep evolutionary relationships.

The work was laborious. This was not the era of pushing a sample into a sequencer and receiving a digital tree. Woese and colleagues generated and compared RNA fragments, looking for patterns that could serve as an evolutionary record.

Some methane-producing organisms did not fit within the bacterial branch as expected. Their molecular signatures suggested a lineage as distinct from bacteria as either was from eukaryotes. Woese and George Fox proposed that these organisms represented a separate primary kingdom, later formalized as Archaea and ultimately placed in a three-domain view of life.

The resistance was real and understandable. A new domain did not simply add organisms. It challenged the criteria by which biology had organized living things. Morphology and conventional physiology had made the organisms look bacterial. Molecular sequence revealed a deeper history.

Woese's achievement matters to the ghost genome for two reasons.

First, it helped establish that microbial diversity could be inferred without first culturing every organism. The unit of measurement changed, and an invisible world became legible.

Second, it shows that a new instrument can reveal categories rather than merely objects. Ribosomal RNA did not only identify more microbes. It reorganized the tree itself.

Metabolomics may be approaching an analogous transition. We are not simply finding more compounds to add beneath known pathways. We may be discovering that the useful unit is sometimes a molecular family, a spatial gradient, a transient state, or a cross-kingdom relay rather than one stable metabolite assigned to one organism.

Two earlier observers who arrived before the language

The Galileo Effect appears differently in medicine and genetics.

Ignaz Semmelweis confronted a mortality pattern in nineteenth-century Vienna that should not have existed under the accepted theories of puerperal fever. Women in a physician-led maternity clinic died at much higher rates than women in a clinic staffed by midwives. After a colleague died following an autopsy injury, Semmelweis connected the physician clinic's autopsy work with the deaths and required handwashing in chlorinated lime.

Mortality fell.

The result did not reorganize medicine immediately. Semmelweis lacked germ theory, communicated with increasing anger, and could not provide the mechanism his contemporaries expected. The tragedy is often retold as stupid doctors rejecting obvious evidence. The harder lesson is that an outcome can be dramatic and still remain conceptually homeless when the field lacks a mechanism and the messenger cannot build a bridge.

Barbara McClintock encountered a quieter version. Studying maize, she inferred that genetic elements could move within the genome and alter the expression of neighboring genes. The observation conflicted with the simpler picture of genes as fixed beads on a chromosome. Her experiments were meticulous, but the conceptual language and molecular tools of the period could not easily absorb what she saw. Transposable elements later became foundational biology.

Neither story proves that an unsupported claim is secretly true. They show that observations can arrive before categories.

The modern obligation is to build the bridge faster: preserve the anomaly, develop the method, define the failure condition, and translate the finding into a form the field can test without requiring the observer to become a martyr.

The uncultured majority

The Great Plate Count Anomaly was known long before modern sequencing. Microbiologists could see far more cells under a microscope than they could grow as colonies on standard laboratory media. The discrepancy was not subtle. Nature was full of organisms that the Petri dish failed to recover.

Culture-independent sequencing made the uncultured majority visible in a new way. It exposed entire lineages, genes, and biosynthetic capacities without requiring organisms to accept laboratory conditions.

Metagenomics then produced another surprise. Even when organisms and genes could be detected, large fractions of the sequences had no close functional reference. Biosynthetic gene clusters suggested enormous specialized-metabolite capacity. Yet sequence still did not reveal the complete chemical output of a living community.

The telescope had moved the boundary. It had not finished the map.

Mass spectrometry became a second telescope.

Untargeted experiments routinely produce thousands or tens of thousands of features. Only a fraction can be confidently assigned to structures using available reference libraries and standards. Modern computational tools have improved annotation, but the central fact remains: the measured chemical world is larger than the named chemical world.

This difference is often hidden in a results table. Known metabolites receive names and pathways. Unknowns become feature IDs, supplemental columns, or discarded rows. The instrument is permitted to record the anomaly without allowing the anomaly to affect the ontology.

That is the modern form of looking through the telescope while refusing what it shows.

A telescope with several lenses

The chemical world requires more than one instrument because each method sees a different projection.

High-resolution liquid-chromatography mass spectrometry separates compounds over time and measures ions with exquisite mass accuracy. Tandem mass spectrometry fragments selected ions, creating patterns that can suggest substructures and relationships.

Molecular networking connects spectra that fragment similarly. A feature without a library match no longer has to remain an isolated number; it can sit within a family, perhaps beside a known compound. The network does not prove structure. It organizes uncertainty and reveals that the unknown is patterned rather than random.

Spatial mass spectrometry preserves location. A laser or other probe samples points across tissue, biofilm, root, or microbial colony, producing a molecular map. A feature that looks unremarkable in a homogenized average may form a sharp boundary at a tumor edge, mucus layer, wound, or competing colony.

Stable-isotope tracing follows atoms. A labeled dietary precursor, bile acid, or amino acid enters the system. Mass shifts reveal which downstream products contain the label. Defined microbial communities and host models can identify which organism or enzyme is required at each step.

Single-cell and microfluidic methods reduce the scale of observation. Organoids and tissue models preserve parts of architecture. Nuclear magnetic resonance and synthesis confirm structures. Phenotypic assays ask what the chemistry does before a complete target story exists.

The lenses overlap imperfectly. A strong discovery is often the point where several agree.

This is why the ghost genome cannot be solved by sequencing more deeply, measuring more peaks, or generating more structures in isolation. The object is relational. The telescope must preserve enough of the relationship for the signal to remain meaningful.

The Observability Ladder

A modern mass spectrometer can detect a molecular feature without giving us an immediate picture of the molecule. In that respect, the feature resembles the wobble of 51 Pegasi.

The instrument records constraints: mass-to-charge ratio, retention time, isotope pattern, ion mobility, and perhaps fragmentation. Those measurements may show that something physical is present and related to other features. They do not automatically reveal full structure, stereochemistry, producer, location, or function.

The scientific response should not be a choice between declaring a discovery and deleting the unknown. It should be a ladder of increasing observability.

1. Trace: show that the signal exceeds background and survives basic quality controls.
2. Reproducibility: show that it returns across injections, samples, collections, or laboratories.
3. Context: preserve where, when, and under what ecological or physiological conditions it appears.
4. Capture: recover enough material, or create a targeted method, so the feature can be followed through separation.
5. Identity: distinguish one molecule from adducts, fragments, isomers, and contaminants; confirm the candidate with orthogonal chemistry and; where possible, an authentic standard.
6. Causality: show that the confirmed chemistry changes a biological system at plausible exposure and that removing it, its producer, or its pathway changes the effect as predicted.

7. Translation: demonstrate that the mechanism survives the geography, metabolism, safety constraints, manufacturing, and patient variability of the intended use.

The ladder prevents two equal and opposite errors. It prevents us from treating every unexplained peak as profound. It also prevents the existing library from defining reality by excluding anything it cannot name.

Unknown is not a conclusion. It is a coordinate with a confidence level and a next experiment.

Mayor and Queloz did not announce a planet because one spectrum looked unusual. They measured periodicity, tested alternatives, and allowed the inferred object to become more specific as evidence accumulated. Chemical discovery should mature in the same way.

A field advances when enough coordinates reveal an architecture that the old map could not represent.

The observation that remains outside the model

Institutional resistance in modern science rarely looks like someone refusing to examine data. It looks like accepting the measurement while preserving the old categories.

An unknown spectrum is labeled *unannotated* and excluded from the biological story.

A microbial effect is treated as a confounder around a human target rather than part of the target system.

A multi-target natural product is called promiscuous because the assay framework assumes one intended interaction.

A large molecule is called non-drug-like even when molecules with similar properties are approved medicines or endogenous signals.

A community-dependent phenotype is decomposed until the community no longer exists.

The observation enters the database but not the model.

51 Pegasi b could not be understood by placing it in the usual category of a Jupiter-like planet at a Jupiter-like distance. The category itself had to expand. Hot Jupiters required new accounts of formation and migration.

In the same way, the ghost genome may require chemistry to be organized by more than molecular identity. Location, timing, community, conformational state, and production history may be part of what the molecule *is* biologically.

A bile-acid derivative in the intestinal lumen is not equivalent to the same nominal structure in portal blood. A peptide bound to a metal or membrane can present a different surface from the free molecule in solution. A transient product released during phage lysis can create a local pulse that disappears from a bulk sample. A molecular family may act through a pattern of related targets rather than one highest-affinity interaction.

The chemical name remains essential. It may not be the complete unit of function.

A planet of chemical experiments

The new microbial map changes more than taxonomy. It changes the scale of the chemical history we inherit.

Before there were animals, plants, organs, nerves, or immune systems, there was chemistry under selection. The earliest cells had to create boundaries, harvest energy, store information, copy themselves, sense danger, repair damage, and compete for resources. They had to decide when to grow, when to wait, when to cooperate, and when to attack.

Microbes spent billions of years working on those problems before a human receptor existed.

Evolution is often pictured as a sculptor changing bodies. In microbial life, evolution also changes enzymes, pathways, signals, toxins, transporters, and ecological conversations. The result is not a single orderly library. It is a planetary chemical history written through mutation, duplication, recombination, loss, symbiosis, and horizontal gene transfer.

Life repeatedly reuses successful building blocks. Amino acids become peptides. Acetate and malonate become polyketides. Isoprene units become terpenes. Sugars are attached, removed, oxidized, and rearranged. Fatty acids become membranes and signals. Aromatic amino acids become indoles and phenolics. Metals are coordinated. Rings are closed. Stereochemistry is controlled.

Nature does not search every theoretically possible structure. Cells work from available atoms, enzymes, energy, transport, and ecological need. That constraint is not an embarrassment. In an otherwise immense chemical universe, it can be an advantage.

A physically recovered natural molecule has survived a set of filters that a virtual structure has not. A cell or community assembled it. Precursors existed. Enzymes or chemical conditions were capable of producing it. The molecule persisted long enough, in some form, to be detected or recovered. It participated in an ecology that included membranes, water, competing organisms, reactive chemistry, and selection.

None of that proves the molecule is safe or useful to humans. Nature produces toxins, carcinogens, and metabolic liabilities with equal creativity. Evolution optimizes the fitness of the producing system, not the well-being of a future patient.

But biological existence is information.

It is a prior.

Human receptors evolved inside the microbial world

Animals did not develop in chemical isolation. Every mucosal surface evolved in contact with microbes. Immune systems learned to distinguish food, damage, pathogens, symbionts, and self in an environment saturated with microbial products. Pattern-recognition receptors sense conserved microbial structures. G-protein-coupled receptors respond to fatty acids, bile-acid derivatives, and other metabolites. Nuclear receptors integrate dietary, microbial, host-derived, and environmental ligands.

This does not mean every microbial molecule was designed for a human receptor. Evolution has no design brief. It means the receptor landscape and the microbial chemical landscape have been interacting over deep time.

The interaction can be cooperative, antagonistic, or accidental.

Short-chain fatty acids produced through microbial fermentation can nourish colonocytes and influence immune and metabolic pathways.

Microbial transformations of primary bile acids create ligands with different receptor activities. Tryptophan metabolism generates compounds that engage the aryl hydrocarbon receptor and other host pathways. Bacterial N-acyl amides can resemble human signaling lipids and activate human G-protein-coupled receptors.

These connections are experimentally demonstrated. They also represent the named edge of a much larger unknown.

Once 51 Pegasi b was discovered, astronomers did not conclude that every unexplained stellar wobble was a planet. They improved instruments, built catalogs, developed confirmation standards, and learned which signals were produced by stellar activity or systematic error. The result was not credulity. It was a more capable skepticism.

Microbiome chemistry needs the same transition. The existence of known microbial ligands does not make every unknown peak a signal. It makes the search for additional signals rational.

The outside world as an expanded archive

Environmental and human-associated microbiomes are not interchangeable. The human gut has temperature, mucus, host immunity, dietary rhythms, and intimate tissue contact. A marine bloom has light, salinity, grazers, minerals, currents, and different nutrient pressures. Soil, wounds, roots, skin, and lungs each select different chemistry.

Yet the catalytic grammar overlaps. Nonribosomal peptide synthetases, polyketide synthases, ribosomally synthesized and modified peptides, terpenoid pathways, glycosylation systems, proteases, ligases, and tailoring enzymes appear across habitats. Deep ancestry and horizontal gene transfer connect them.

The outside world is therefore not a substitute for the human microbiome. It is an expanded evolutionary archive of structures and strategies that living systems have physically realized.

This distinction matters to drug discovery. A molecule from an aquatic community may never have existed in a human intestine. Its architecture may still reveal a way to bind a conserved protein surface, organize an induced interface, resist proteolysis, cross a membrane, remain localized, or coordinate several targets. Environmental chemistry can be both a source of direct candidates and a teacher of molecular solutions.

The strongest defensible claim is modest: environmental microbiomes provide finite, physical priors drawn from a shared biochemical grammar. The more speculative claim is that some of those priors can help reconstruct or inspire missing coordination chemistry in the human system.

That claim must be tested molecule by molecule.

The broader point is that the new telescope does not merely show us more microbial species. It reveals a planet that has been conducting chemical experiments for billions of years while medicine sampled only the fraction willing to grow in our flasks and fit our rules.

Productive heresy is an experimental discipline

Outsiders are not automatically right. Most unconventional ideas are wrong, and some are dangerous. The value of the outsider is not superior intelligence. It is different exposure to assumptions and incentives.

An insider sees a claim against decades of failed claims in the same field. An outsider may see the field's premise as one option among several.

An insider knows how difficult the accepted problem is. An outsider asks whether it is the right problem.

An insider has specialized language, methods, and reputation. An outsider has fewer sunk costs and more permission to cross boundaries.

This is the heretic's advantage: not freedom from error, but freedom from one particular map.

The advantage becomes useful only when paired with discipline.

Name the established facts. Separate observation from interpretation. Define what result would weaken the hypothesis. Use the old experts to identify hidden confounders. Build the instrument. Invite independent replication. Do not turn rejection into evidence of correctness.

Galileo did not win because he was an outsider. Mayor and Queloz did not win because a hot Jupiter made a compelling headline. Woese did not change biology because the three-domain tree was aesthetically pleasing.

The moons were there. The wobble repeated. The RNA sequences separated.

The hidden chemistry of life will become real in the same way: molecule by molecule, spectrum by spectrum, assay by assay, across laboratories that do not share the same incentives.

The next story follows one of those molecules from an isolated island to a corporate laboratory, into a household freezer, and back into medicine. It is the story of an outlaw compound that violated the industry's rules before the rules knew how to describe it.

PART II

Nature's Chemical Memory

Three stories about molecules and communities that changed the meaning of biological function.

• • •

CHAPTER 4

The Molecule in the Freezer

The Outlaw That Survived Its Corporate Death

STORY 4

THE MOLECULE IN THE FREEZER

The Outlaw That Survived Its Corporate Death

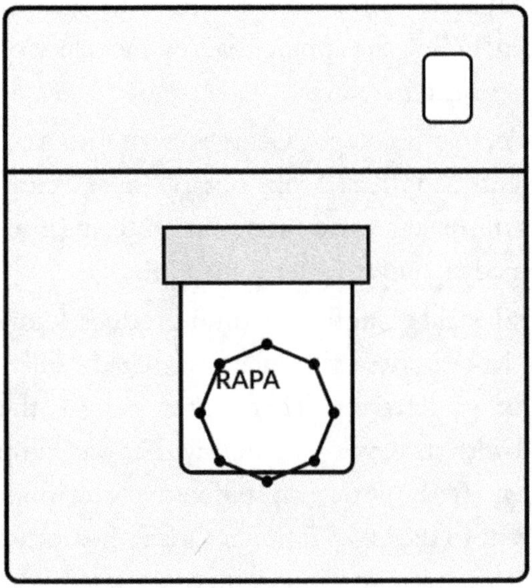

THE TURNING POINT

The institution stopped. The physical evidence remained alive.

In 1964, a Canadian naval vessel left Halifax for one of the most isolated inhabited places on Earth.

The expedition aboard HMCS *Cape Scott* was headed to Rapa Nui, Easter Island, before a planned airport and increasing outside contact altered the island's population and biosphere. The mission was broad: document health, ecology, and microbial life in a place that had long seemed geographically self-contained.

Among the collected materials were soil samples.

The original scientific question was not mTOR, transplantation, cancer, aging, or any of the subjects that would later attach themselves to the story. Researchers were interested in the island's microorganisms and the possibility that local soil chemistry might help explain observations about infection, including the apparently low incidence of tetanus among people who often walked barefoot.

Back in Canada, microbiologist Georges Nogrady's samples reached the Ayerst laboratory in Montreal. From one of them, researchers isolated a *Streptomyces* organism that produced an antifungal compound. The molecule was named rapamycin after Rapa Nui.

At first, it looked like another natural-product lead in an era when pharmaceutical laboratories routinely searched soil organisms for antibiotics and other activities. Then Suren Sehgal, the scientist most closely associated with its development, saw that the compound did more than inhibit fungi. It showed antiproliferative and immunosuppressive effects. Samples sent to the U.S. National Cancer Institute drew interest.

The molecule was becoming more important at precisely the wrong corporate moment.

Ayerst closed the Montreal research facility during consolidation. Programs ended. Samples were supposed to be discarded. Rapamycin had no approved indication, no established mechanism, no commercial product, and no organizational home.

Sehgal took cultures and material home and placed them in his freezer.

The image is irresistible because it compresses an entire theory of innovation into one domestic appliance. Corporate priorities had declared the program dead. The molecule remained physically alive as a possibility, stored beside the ordinary contents of a household.

Years later, the work was revived. Rapamycin's mechanism became clearer. The compound binds the intracellular protein FKBP12; the resulting complex engages mTOR, a central regulator of growth, metabolism, and cell-state decisions. The drug ultimately entered medicine as sirolimus, approved in the United States in 1999 as an immunosuppressant, and became a tool that reshaped basic research across cancer, metabolism, immunology, transplantation, and aging biology.

The freezer did not prove that every abandoned compound deserves rescue. Most abandoned compounds are abandoned for good reasons. The story matters because rapamycin violated several categories before those categories knew how to describe it.

It was too complex to look like the ideal small molecule. Its mechanism was relational rather than a simple one-drug/one-pocket interaction. Its biological effects crossed systems. Its value emerged over decades and through changing questions. The molecule did not fit the dashboard of the organization that first encountered it.

It survived because one scientist recognized that the map might be temporary while the molecule was real.

What the freezer preserved

A freezer preserves more than material. It preserves optionality.

When the Montreal laboratory closed, rapamycin's future depended on scientific questions that had not yet matured. Its antifungal activity was interesting but not obviously superior to existing opportunities. Its immunosuppressive effect could look like a liability in an anti-infective program. Its antiproliferative activity required a development logic, assays, and disease hypotheses beyond the original search.

The same observation can change category as knowledge changes.

A compound that suppresses immune function may be unsuitable for infection and valuable in transplantation. A molecule that slows cell growth may be toxic in one setting and useful in cancer. An effect on wound healing may be an adverse event for one indication and a mechanistic clue in another. The molecule remains the same while the clinical question changes.

This is one reason natural-product programs are vulnerable to corporate transitions. Their value is often option-like rather than immediately asset-like. A well-defined development candidate can be evaluated against a target product profile. An unusual molecule with several activities requires an organization to preserve uncertainty long enough to discover which question makes it valuable.

Corporate governance is built to compare opportunities. It must allocate finite capital, stop weak programs, and avoid keeping every interesting compound alive. The discipline is necessary. The risk arises when the same scorecard is applied to a known asset and a new mechanism.

Rapamycin's home freezer became a temporary institution with a different metric: the material was unusual enough not to destroy.

When Ayerst later reconstituted research and the program returned, the preserved culture allowed the scientific opportunity to reenter the formal system. The story should not be read as an invitation for employees to remove company property. It is a historical illustration of how fragile physical evidence can be when an institution changes direction.

A digital record saying *interesting activity observed* would not have reproduced the organism.

The material mattered.

When a laboratory closes, a future closes with it

A research laboratory contains more than projects. It contains material histories.

A culture tube can preserve a lineage that no catalog fully describes. A notebook can contain the failed condition that explains why the successful one mattered. A technician remembers that one fraction changed color when exposed to air. A chemist knows which chromatographic shoulder carried the activity even though the official report shows only the final peak. A freezer accumulates organisms, extracts, standards, intermediates, and anomalous samples whose value may not yet be legible.

When a laboratory closes, that memory is disassembled.

Some of it is transferred. Some is archived. Some is destroyed because storage costs money and ownership creates obligations. Some disappears because no one remaining understands why the material was retained. The loss is not always a mistake. Institutions cannot preserve everything indefinitely. But natural-product discovery is unusually exposed because the physical source may be difficult to recreate.

A synthetic series can sometimes be reconstructed from a structure and a route. A microbial isolate collected from a specific place at a specific time may no longer exist in the same ecological state. A fraction derived from an uncultured community may contain a chemistry that was never assigned to a producer. The notebook may say *activity in fraction 47*, but if fraction 47 is discarded, the evidence becomes an anecdote.

This is why sample governance is part of scientific strategy.

A mature discovery organization needs explicit categories for physical optionality:

- material essential to an active asset;
- material linked to a reproducible but unexplained activity;
- material representing a unique source, time, or ecology;
- material whose value has been superseded by a confirmed structure and sustainable resupply;
- material retained only by habit and safe to retire.

The decision should not be made by freezer space alone, nor should every vial acquire the emotional status of rapamycin. The point is to preserve the connection between the reason a sample mattered and the cost of recreating it.

Digital data make that judgment better when they remain linked to the physical object. A spectrum without a vial may be unconfirmable. A vial without metadata may be uninterpretable. Together they become a recoverable scientific option.

Sehgal's household freezer is memorable because it was informal. The larger lesson is institutional: discovery systems need a legitimate way to protect anomalous material from being erased by a change in corporate calendar.

The visual style of medicinal chemistry

Medicinal chemistry has a visual style.

Open a catalog of conventional small-molecule drugs and many structures look as though they were drafted on graph paper: aromatic rings, short linkers, familiar functional groups, relatively few stereocenters. The structures are not simple in any ordinary sense. Their design may represent years of ingenious optimization. But much of twentieth-century screening favored compounds that were easier to synthesize, store, measure, and fit into well-defined target pockets.

Natural products often look as though a different civilization made them.

They bend out of the page. They contain multiple rings, stereocenters, sugars, unusual amino acids, lipid tails, internal hydrogen bonds, and dense patterns of oxygen and nitrogen. Some are macrocycles - large rings capable of adopting several related conformations. Some are peptides assembled without the ribosome. Some combine peptide, polyketide, carbohydrate, and lipid features in one architecture.

The industry learned to call many of these structures *non-drug-like.*

The molecules did not notice.

Cyclosporine is orally active despite a molecular mass far beyond the conventional Rule-of-Five range. Tacrolimus, sirolimus, vancomycin, paclitaxel, and other clinically important natural products violate one or more familiar property rules. Their success does not abolish those rules. It reveals the domain in which the rules were derived.

What Lipinski actually gave the field

Christopher Lipinski and colleagues examined properties associated with poor absorption or permeation in orally administered drugs. The resulting Rule of Five became one of medicinal chemistry's most useful heuristics. Compounds are more likely to face oral-absorption problems when molecular weight, lipophilicity, or hydrogen-bond donors and acceptors exceed certain ranges.

A heuristic is not a law.

Lipinski's work helped teams avoid the large, greasy, insoluble synthetic compounds that optimization programs can produce. It saved time and improved judgment. The error came later, when software filters, screening libraries, investment habits, and organizational shorthand turned a probability guide into a border defining what a drug was allowed to look like.

The border was efficient. It also excluded chemical neighborhoods in which nature had developed strategies that simple descriptors could not capture.

The field now uses *beyond Rule of Five*, or bRo5, for molecules that operate outside conventional property space. The category includes many macrocycles, constrained peptides, and other larger compounds. It is not one coherent class, and size alone provides no advantage. Most large molecules do not become oral drugs.

The successful ones use strategies.

One is conformational restriction. A flexible linear molecule pays an energetic cost when it must freeze into the precise shape required for binding. Ring closure can preorganize the structure and present functional groups in a target-ready geometry.

Another is chameleonicity. Some macrocycles expose different surfaces in different environments. In water, polar groups may face outward. Near a membrane, internal hydrogen bonds can shield those polar atoms and create a more lipophilic exterior. The molecule behaves less like a rigid key than a folding tool.

A third is local action. A molecule need not enter systemic circulation to be useful. It can remain in the gut, skin, lung, eye, joint, or wound. Properties that look like failures under a systemic-delivery model can become tools for geofencing exposure.

A fourth is transporter participation. Some large molecules interact with uptake or efflux systems. This is compound-specific, not a universal privilege of natural products, but it reminds us that passive diffusion is not the only route through biology.

A fifth is relational binding. A larger, three-dimensional structure can engage broad or shallow surfaces, contact a pocket and neighboring protein surface, or stabilize an interface that did not exist before the molecule arrived.

Rapamycin is the classic lesson.

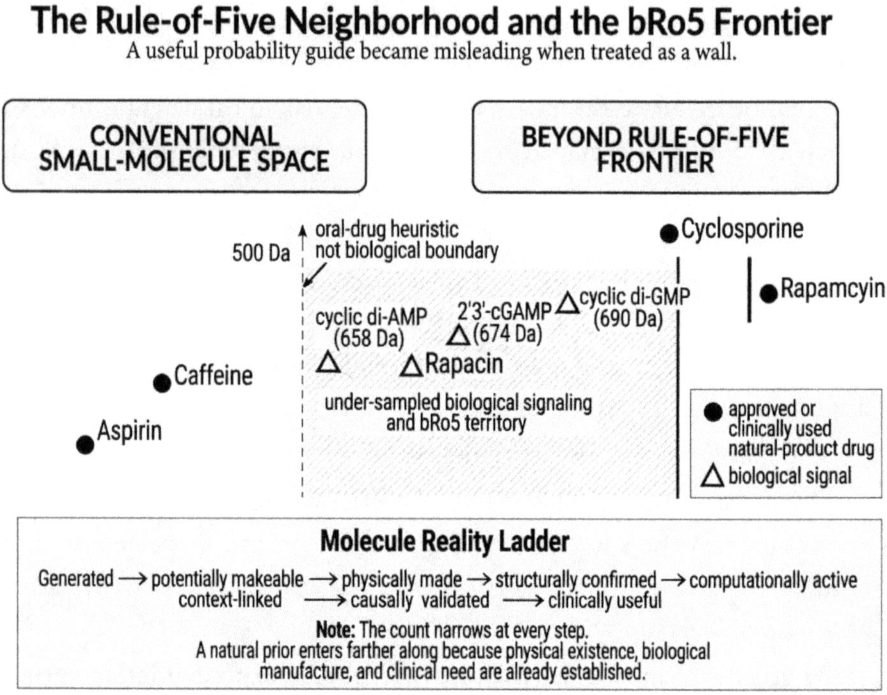

Figure 5. The conventional Rule-of-Five neighborhood and the beyond-Rule-of-Five (bRo5) frontier. The boundary is a guide, not a wall.

Biology Did Not Stop at 500 Daltons

The five-hundred-dalton boundary was developed as a practical guide to the probability of oral absorption for many conventional drugs. It was never a boundary around biological signaling. Living systems use chemical messages across a wide continuum—from ions, gases, and small metabolites to lipids, nucleotides, peptides, macrocycles, and molecular assemblies.

Consider the cyclic dinucleotides. Cyclic di-AMP has a molecular weight of approximately 658 daltons. The mammalian immune second messenger 2′3′-cGAMP is approximately 674 daltons. The bacterial second messenger cyclic di-GMP is approximately 690 daltons. These are real control molecules. They help cells sense danger, alter gene programs, organize growth, and change collective behavior. They are not conventional oral drug candidates; they are highly polar signals whose biology depends on production, transport, degradation, and compartmentalization. Their importance is not that they violate a drug rule. It is that biology never agreed to the rule.

The region between conventional small molecules and proteins is therefore not empty. It contains discrete, information-rich chemistry that standard libraries, assays, extraction methods, and predictive models may underrepresent. Macrocycles and constrained structures in this region can present larger three-dimensional surfaces, create or stabilize protein interfaces, coordinate several targets, or remain localized at a biological boundary.

Approximately 750 daltons is useful in this book as a discovery frontier, not as a magic number or an upper limit. Many important signals are smaller, and some are considerably larger. The lesson is not that every molecule in this expanded range should be generated and screened. That would worsen the search-space problem. The lesson is that physical biological priors give us a disciplined way to enter a broader chemical territory without searching it blindly.

How large molecules cross the border

The physics of beyond-Rule-of-Five molecules helps explain why a flat descriptor can mislead.

A molecule in solution does not possess one permanent three-dimensional shape. It occupies an ensemble of conformations. Binding selects and stabilizes some members of that ensemble. Membrane passage may favor another. The energetic cost of moving among states affects both potency and permeability.

Macrocyclization reduces some flexibility by connecting the ends or internal positions of a scaffold. The ring can preorganize key functional groups and reduce the entropy lost during binding. At the same time, a macrocycle may retain enough controlled flexibility to adapt to different environments.

Intramolecular hydrogen bonds are particularly important. A polar atom that forms a hydrogen bond with another group inside the molecule can become less exposed to water. The effective polar surface seen by a membrane may be lower than a two-dimensional count suggests. Once the molecule reaches a binding site, the internal bond can open and the same polar groups can interact with the target.

This is chameleonicity in physical terms: not magical shape-shifting, but an environment-dependent redistribution of surface area and conformation.

The property is difficult to predict from a single structure. Solvent, pH, membrane composition, binding partners, and stereochemistry matter. Two analogs with nearly identical formulas can behave differently because one can form a productive internal hydrogen-bond network and the other cannot.

Nature is unusually good at building such structures because enzymes control stereochemistry and sequence. A nonribosomal peptide synthetase can incorporate unusual amino acids, methylate amides, add lipid groups, and close rings. Tailoring enzymes can oxidize, glycosylate, halogenate, or cross-link the scaffold. The resulting molecule is a three-dimensional device, not merely a collection of fragments satisfying descriptor thresholds.

This does not make bRo5 chemistry easy. Synthesis, solubility, permeability, formulation, metabolism, and analytical control can all be difficult. The point is that difficulty should be measured rather than assumed from a border designed around a different population of molecules.

Cyclosporine's impossible passport

Cyclosporine offers another way to understand how a large natural molecule crosses a border that a simple descriptor says should be closed.

On paper, cyclosporine looks like a poor oral drug. It is large, cyclic, rich in amide bonds, and far beyond the conventional molecular-weight boundary. A linear peptide with similar size and polarity would usually be degraded, poorly absorbed, or both.

Cyclosporine behaves differently because the structure is not static.

In a polar environment, it can expose groups that support solubility. In a less polar environment, internal hydrogen bonds and conformational changes can hide some of that polarity from the membrane. The molecule effectively presents different passports at different checkpoints.

This behavior is not magic and not unique to cyclosporine. It is a physical consequence of shape, intramolecular interactions, solvent, and energy. It also explains why two-dimensional descriptors can mislead when used without conformational information.

A molecular-weight value is real. A hydrogen-bond count is real. A calculated polar surface area is real. But a living membrane does not encounter a spreadsheet row. It encounters a fluctuating three-dimensional object in a particular environment.

For AI-enabled drug design, this distinction is consequential. Models trained mainly on smaller, relatively rigid molecules may predict properties from two-dimensional representations with impressive average performance. The same model can become unreliable in the bRo5 frontier, where permeability may depend on rare conformations, solvent-dependent hydrogen bonding, transporter interactions, aggregation, or slow interconversion among states.

The answer is not to abandon prediction. It is to improve the ground truth.

Measurements of conformational ensembles, permeability, efflux, solubility, binding, and intracellular exposure can teach models which large architectures succeed and why. Natural products supply unusually rich examples because evolution and biosynthesis have already explored ways to stabilize complex three-dimensional structures.

They also supply failures. Many macrocycles are insoluble, unstable, impermeable, difficult to manufacture, or toxic. Those negative examples are just as valuable if they remain visible.

The goal is not a new rule saying that large molecules are better. It is a more accurate question: *Which structural strategies allow a particular large molecule to do the job required in the geography where it must act?*

That is a design problem, not a boundary violation.

A molecule that creates a target

The simplest pharmacology diagram contains a ligand and a target. The molecule enters a pocket and activates or blocks the protein.

Rapamycin does something more interesting. It first binds FKBP12. The combined surface of drug and protein then binds mTOR. The functional target is not merely a preexisting pocket waiting for one small molecule. The drug helps create a new protein-protein relationship.

This mechanism belongs to the broader world of molecular glues and induced proximity, where a compound reorganizes which biological components can interact. It challenges the common assumption that a drug's job is always to occupy one site on one protein as selectively as possible.

The ghost-genome hypothesis extends that lesson. If microbial communities generate relational molecules - ligands, glues, fragments, conjugates, metal-bound states, and transient complexes - then some unknown chemistry may coordinate biology by changing which components can talk to one another.

That is not a claim that every natural product is a molecular glue. It is a reminder that our screening ontology influences which mechanisms we can discover. A library optimized around isolated enzyme inhibition may miss a compound whose function requires a partner protein, membrane, community metabolite, or spatial boundary.

The molecule in Sehgal's freezer was not only a drug candidate. It was evidence that biological chemistry could use a mechanism the prevailing small-molecule picture underrepresented.

Chords, not notes

A piano key makes a note. A chord creates a relationship.

For much of modern pharmacology, the ideal drug was imagined as a precise key for one molecular lock. Paul Ehrlich's *magic bullet* became the enduring image: a therapy that strikes the cause of disease without harming the patient. The aspiration was humane and productive. Selectivity matters. Targeted medicines have transformed cancer, infection, inflammation, and rare disease.

The metaphor becomes dangerous when precision is equated with singularity.

Complex diseases rarely reside in one protein. They emerge from networks with redundancy, compensation, feedback, tissue differences, and changing states. Blocking one node can help, do nothing, or trigger an alternative route. The same target can be harmful in diseased tissue and essential elsewhere.

A molecule that touches several targets is therefore not automatically imprecise. It may be badly promiscuous. It may also be appropriately polypharmacological.

Promiscuity describes broad or poorly controlled interactions. A hydrophobic compound that sticks to many proteins, blocks hERG, inhibits several metabolic enzymes, and damages membranes has many targets but no coherent therapeutic logic.

Polypharmacology describes intentional or functionally meaningful modulation of multiple targets. The targets may sit in one family, parallel pathways, or complementary points in a disease network.

The difference is pattern, dose, location, and time.

A multi-kinase cancer drug can be effective because it suppresses several growth and angiogenesis pathways. It can also be toxic because those kinases support normal tissues. A psychiatric medicine may derive benefit from a blend of receptor and transporter activities while generating side effects through another subset. A microbial metabolite may engage a cell-surface receptor and a nuclear pathway at concentrations that produce a coordinated physiological shift.

The molecule is a chord. Whether the chord resolves or grates depends on which notes, in which tissues, at what volume, and for how long.

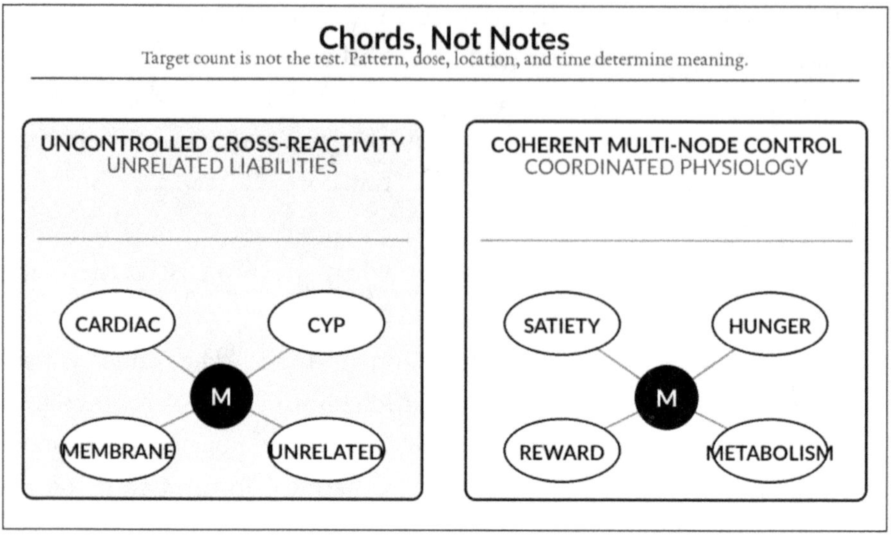

Figure 6. Chords, not notes: uncontrolled cross-reactivity versus coherent multi-node control.

The Master-Key Hypothesis - Properly Qualified

The materials that informed this book contain a provocative idea: one evolution-shaped scaffold may sometimes influence several related physiological controls because the pattern served a coherent ecological function before anyone imagined it as a medicine.

The phrase master key is useful only if we resist its most dramatic meaning. A molecule is unlikely to be a universal key that opens unrelated biological doors at will. A better image is a change key cut for a family of related locks. Its three-dimensional shape, charge pattern, and conformational behavior may allow it to engage several members of a physiological network with different strengths.

In appetite regulation, for example, satiety, hunger initiation, reward, stress, and energy expenditure are not separate machines. They are overlapping controls. An evolution-shaped molecule that modestly changes several nodes could produce a more coherent state shift than a compound that saturates one receptor. The same architecture could apply to inflammatory resolution, tissue repair, microbial competition, or stress resilience.

This is a hypothesis, not an explanation already earned by a target-prediction table. Computational convergence can identify an interesting pattern. It cannot precisely establish agonism, antagonism, free exposure, tissue distribution, necessity, safety, or evolutionary purpose.

The stronger evolutionary story - that one ancient microbial signal was directly transferred into animal genomes and that several modern receptors co-evolved around it - remains conjectural. Horizontal gene transfer is common among microbes, and host-microbial co-evolution is real, but a specific gene-transfer route or receptor ancestry must be demonstrated rather than narrated.

A disciplined version of the master-key hypothesis makes predictions. The target pattern should be organized rather than hydrophobic noise. Close analogs should remove or rebalance particular activities in ways that change the integrated phenotype. The active exposures should be plausible in the relevant tissue. Comparative biology should reveal a credible ecological route. And the useful chord should separate from safety liabilities during optimization.

If those tests fail, the scaffold may still be a useful single-target ligand or an instructive natural product. If they succeed, it would support a larger idea: evolution can supply not only potent keys, but keys whose shape carries a systems-level logic that medicinal chemistry can refine. Many successes may lead to de novo (from scratch) polypharmacology AI toolkit.

Why biology uses multi-node control

Biological systems are robust because they do not depend on one switch.

Blood glucose is regulated through insulin, glucagon, incretins, neural input, hepatic metabolism, muscle uptake, adipose signaling, circadian timing, and behavior. Inflammation includes initiation, amplification, containment, repair, and active resolution. Appetite integrates sensory inputs, gastric distension, gut hormones, reward pathways, stress, memory, sleep, and microbial metabolites.

A single-target intervention can be powerful when one node is rate-limiting. In other cases, it is like steering a ship by moving one rope.

Microbiome-derived molecules often enter these systems as modulators rather than master switches. Short-chain fatty acids can serve as substrates, receptor ligands, and epigenetic modifiers. Bile acids act as detergents, metabolic intermediates, antimicrobial pressures, and signals through nuclear and membrane receptors. Tryptophan derivatives can influence barrier, immune, and neural pathways through several mechanisms.

These examples do not prove that every microbial metabolite is a carefully tuned homeostatic agent. They demonstrate that multifunctionality is normal biology, not an aberration that appears only when a drug misbehaves.

This changes the discovery question. Instead of asking only, *What is the molecule's strongest target?* we must ask, *What pattern of interactions occurs at the exposure the tissue actually experiences?*

The target panel removes geography. The living organism restores it.

Phenotype remembers what the target assay removes

Natural products and phenotypic discovery have historically fit together because both begin with function.

Before target-based screening became dominant, many medicines were recognized because a substance killed a pathogen, changed blood pressure, altered behavior, suppressed rejection, or improved an animal phenotype. The mechanism was often discovered later.

Phenotypic screening has real weaknesses. An effect can be difficult to deconvolute. The assay may be fragile. Optimization without a clear target can become slow. A mixture may lose activity when separated because several components are required.

But phenotype preserves relationships that disappear when biology is reduced to one purified protein.

Modern tools make the older strategy more rigorous. Chemoproteomics can identify engaged proteins across cells. CRISPR screens can reveal genes required for a response. Transcriptomics and imaging can map state changes. Target-deconvolution methods can distinguish direct binding from downstream effects. Close analogs can separate activities.

This creates a better ambition than one target per drug: designed coherence.

A multi-target natural product can be optimized as a *unimolecular cocktail* when its activity profile is mechanistically useful. One structure carries several related actions with one pharmacokinetic profile. The advantage is coordination. The disadvantage is that the ratio of activities is fixed and small chemical changes can improve one target while damaging another.

The development program must identify which notes are necessary, which are tolerated, and which create toxicity. Target count alone says nothing.

Rapamycin again provides a useful contrast. Its central effect does not arise from random binding across many proteins. It arises from a precise sequence of interactions that creates a composite interface. The molecule looks complex, but the mechanism is organized.

The ghost-genome opportunity is not to celebrate promiscuity. It is to search for organized biological patterns that conventional single-target libraries underrepresent.

Nature is not safe

The romantic version of natural-products discovery contains a fatal error: because nature made the molecule, the molecule is safer or wiser than human chemistry.

Nature produced botulinum toxin, aflatoxin, ricin, amatoxins, and countless cytotoxic defenses. Microbial products can damage liver, kidney, nerve, marrow, and mitochondria. Evolution optimizes the reproductive success of the producing system, not the safety of a patient receiving a purified dose.

Yet the opposite claim - that complex natural products are inherently dirtier than rational synthetic drugs - is also misleading.

Many toxicities arise from context mismatch. A local ecological signal can become toxic when delivered systemically. A microbial defense compound may hit mitochondria because mitochondria retain their bacterial ancestry. A molecule whose producer carries resistance or sequestration mechanisms may be dangerous to organisms without them.

The right questions are not *natural or synthetic?* They are:

- What ecology produced the molecule?
- At what concentration and location did selection operate?
- Which human homologs or conserved systems can it affect?
- What metabolites will the patient produce?
- Can chemistry preserve the valuable architecture while reducing liabilities?

This is where semisynthesis earns its place. Many important medicines are partnerships between nature and human design. Nature supplies a privileged scaffold. Chemists improve potency, exposure, selectivity, stability, formulation, or manufacturing. A natural toxin can become a useful medicine only after dose, delivery, selectivity, and risk are brought under control.

Even when the outlaw never becomes a product, it can remain a teacher.

The record nature has already built

The long history of approved medicines shows that natural products and their derivatives remain deeply represented, especially in infectious disease and oncology. The record is sometimes used carelessly, as though historical success guarantees future efficiency. It does not.

Classic natural-product discovery benefited from tractable organisms, repeated screening, and a period when many high-value mechanisms were unoccupied. Rediscovering known compounds became common. Isolation and structure elucidation were slow. Combinatorial chemistry and target-based screening offered speed, standardization, and intellectual clarity.

The industry did not abandon nature because everyone forgot its value. It encountered an access and productivity problem.

The modern opportunity is different. Culture-independent genomics reveals biosynthetic capacity. High-resolution mass spectrometry and molecular networking organize unknown families. Spatial methods preserve location. Automated purification, synthesis, and AI can prioritize follow-up. Industrial-scale collection can move low-abundance features into research quantity.

The question is whether these tools can create a new natural-products productivity curve rather than reproduce the old rediscovery problem at greater expense.

Engineers improve mature systems incrementally until a new architecture changes the cost curve.

Audio recording offers a familiar illustration. Cylinders gave way to magnetic tape, compact discs, solid-state storage, and cloud distribution. The progress was not one continuous optimization of the phonograph; several constraints collapsed in steps when the underlying medium changed.

The accompanying chart is an illustrative analogy rather than an audited economic series. Its purpose is to show how a change in architecture can collapse a constraint that incremental improvement had treated as durable.

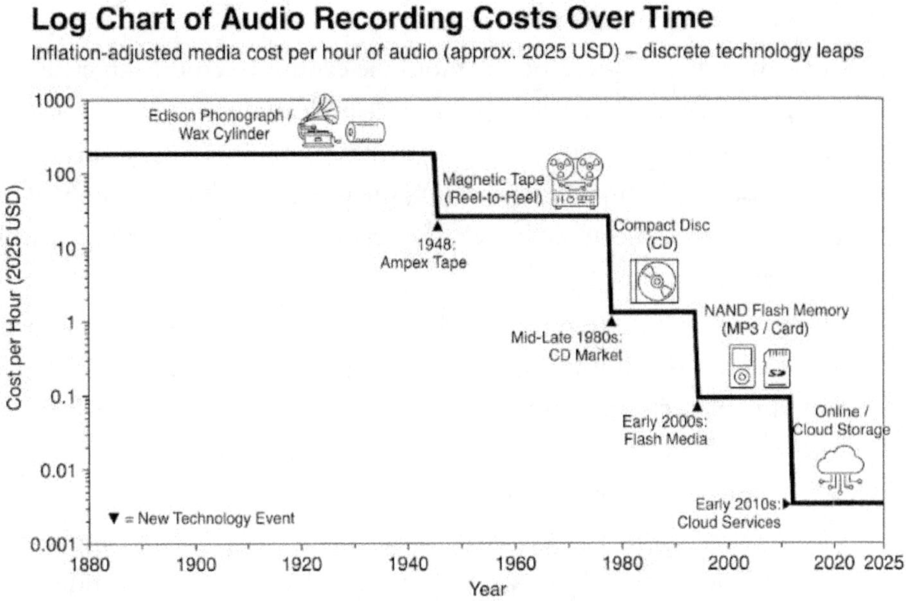

Figure 7. Architectural changes in audio recording. Illustrative, approximate log-scale comparison of media cost per hour; not an audited economic series.

Whether the industry's turn away from natural products improved or weakened long-run discovery productivity is an empirical question.

That is a portfolio question, not a story question. It requires prospective measurement of novelty, confirmatory activity, resupply, optimization, safety, and cost. Famous successes should inspire the experiment, not substitute for it.

Abandonment is a scientific claim

Stopping a program is unavoidable. Treating the stop as intellectually neutral is not.

Every termination embeds a claim about the cause of failure.

The molecule was inactive. The activity was not reproducible. Exposure was inadequate. Toxicity was inseparable from efficacy. Supply was impossible. The mechanism was commercially unattractive. A better competitor arrived. The organization changed strategy. The required experiment cost more than the remaining option value.

Those explanations are not equivalent.

If a program stops because the molecule cannot reach the target tissue, the scaffold may still be useful in a local indication. If it stops because manufacturing is difficult, advances in synthesis or fermentation may alter the conclusion. If a target loses fashion, the phenotype may remain real. If a company exits a therapeutic area, the scientific asset has not necessarily changed at all.

A learning organization records the reason for abandonment in a form that future teams can interrogate. It distinguishes biological failure from organizational decision. It preserves the negative data needed to prevent rediscovery of the same dead end. And it identifies which assumptions would have to change before reconsideration became rational.

Rapamycin's history is an extreme case, but the principle applies to modern AI portfolios. When a model proposes thousands of candidates, most will be rejected. If the rejection is stored only as a binary label, the system loses the explanation. The next model may generate the same liability in a slightly different costume.

The freezer and the failure database perform the same strategic function.

They preserve the option to ask a better question later.

The freezer as a data archive

Public bioactivity databases are shaped by what researchers chose and were able to test. They contain enormous value, but they are not random samples of chemical space. Conventional medicinal chemistry is overrepresented because it has been historically overproduced and easier to standardize. Natural products are underrepresented because isolation, structure elucidation, and synthesis can be difficult. Context-dependent metabolites are underrepresented because many were never purified.

An artificial-intelligence model trained on that record can be technically excellent and still inherit the sampling history.

The model does not know what the system never measured.

Sehgal's freezer was, in one sense, a primitive biological database. It preserved a physical record that the corporate database had decided no longer deserved active resources. When the scientific questions changed, the material could be reopened. A spreadsheet entry or predicted structure could not have substituted for the living producer and stored compound.

The modern equivalent is a traceable physical library linked to spectra, habitat, fraction, biological activity, negative results, structural confidence, and resupply status. The value is not only the molecule. It is the relationship between material and evidence.

Rapamycin's long journey asks a question the rest of this book will keep returning to:

How many valuable structures have disappeared because the organization evaluating them had the wrong question, the wrong assay, the wrong property filter, or no freezer?

The next story moves from one molecule to an entire community. It begins with twins who shared a genome but not a body type, and with mice whose metabolism changed when they received the invisible inheritance carried in a human gut.

CHAPTER 5

The Twins Who Transferred a Metabolism

When a Community Changed the Host

STORY 5

THE TWINS WHO TRANSFERRED A METABOLISM

When a Community Changed the Host

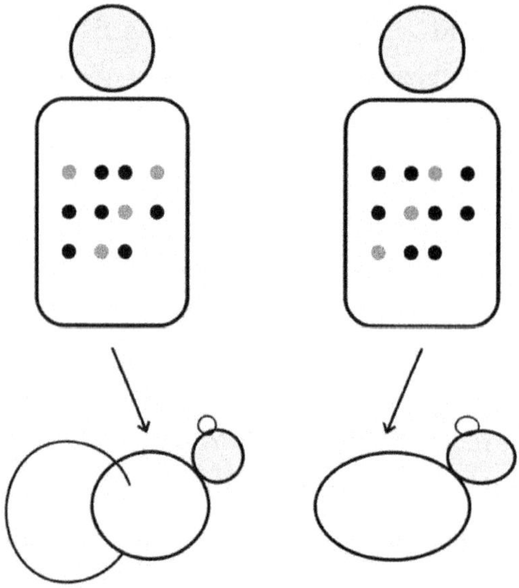

SAME GENES. DIFFERENT COMMUNITY OUTPUT.

THE TURNING POINT

The phenotype moved with ecology - and changed again with diet.

Identical twins are one of biology's most useful natural experiments. They begin with nearly the same genome, yet time, environment, development, illness, behavior, and chance can pull their lives in different directions.

Jeffrey Gordon's group at Washington University studied adult female twin pairs in which one sister was obese and the other was lean. The discordance created a provocative question. If the twins' gut microbial communities were transferred into animals without their own microbes, would any part of the metabolic difference travel with the community?

The recipients were germ-free mice - animals raised in highly controlled conditions without detectable microorganisms. Researchers transplanted fecal microbial communities from the human donors into separate groups of mice and controlled the diets the animals received.

The result was not a metaphor.

Mice receiving microbiota from the obese co-twins developed greater total body mass, increased fat mass, and obesity-associated metabolic features compared with mice receiving communities from the lean co-twins.

The human genome had not been transferred. The human diet had not been copied in full. The mice remained mice. Yet a community of microorganisms carried enough functional information to alter the host's metabolic trajectory.

Then the experiment became more interesting.

Mice harboring an obesity-associated community were housed with mice carrying the lean co-twin's community. Mice eat one another's feces, allowing microbes to move between cage mates. Under the right dietary conditions, members of the lean community invaded the obesity-associated community and prevented the expected increase in adiposity.

The rescue was not universal. Diet mattered. Particular members of the Bacteroidetes were associated with invasion, and a diet rich in certain plant components supported the effect more readily than a diet high in saturated fat and low in fruits and vegetables.

The experiment did not show that scientists had transferred human obesity into mice as a complete disease. It did not prove that one microbial species determines body weight, that a lean person's stool is a treatment, or that genetics, appetite, social conditions, and behavior are secondary.

It showed causal capacity.

A microbial community could transmit aspects of metabolic phenotype. Another community could modify that outcome. The host, community, and diet formed one experiment.

The body was no longer an organism with passengers.

It was a habitat with a metabolism shared across kingdoms.

The experiment was designed to separate inheritance from ecology

The strength of the twin study came from what the researchers controlled and what they allowed to vary.

Human observational studies can show that obesity is associated with a different microbiome, but the direction of causality is difficult to resolve. Diet changes the microbiome. Obesity changes physiology. Medications, activity, age, and socioeconomic conditions can influence both. A microbial pattern may be a consequence rather than a cause.

Germ-free mice provided a recipient whose microbial starting state was known. Transplanting communities from discordant twins allowed the investigators to compare microbiota from people with similar genetic backgrounds while preserving the natural complexity of each community. Controlled diets then tested whether the microbial effect depended on what the host was fed.

The result was not a single *obesity bacterium*. The communities differed in function and ecological behavior. When the lean-associated community invaded the other mice, the change depended on diet and on which organisms could establish themselves.

This is important because the popular imagination prefers one culprit and one hero. One microbe causes obesity; another cures it. The experiment showed something more demanding: the phenotype emerged from a network whose behavior depended on ecological access to substrate.

The mice also remind us that causal capacity is not direct clinical translation. Germ-free animals have unusual immune and metabolic development. Human eating behavior, social environment, genetics, and long-term physiology are not reproduced by a mouse cage. The study established that microbiota can carry a modifiable metabolic influence. It did not estimate how much of human obesity is microbial or identify a universal intervention.

The right response is neither hype nor dismissal. It is to ask which chemical functions moved with the community.

Inside the gnotobiotic room

The mice in the twin experiment did not begin as miniature people with ordinary microbial histories. They were germ-free.

That fact is the power of the model and its limitation.

A germ-free mouse is delivered and maintained in an environment designed to exclude microorganisms. Food, bedding, water, cages, and handling pass through controlled systems. The animal develops without the microbial exposures that shape a conventional immune system, intestinal architecture, metabolism, and behavior.

The absence creates a biological blank that does not exist in normal human life.

Into that blank, investigators can introduce a defined community or a human donor's microbiota and ask what changes. The experiment is not a perfect simulation of the donor. It is a causal instrument. By holding host genetics, housing, and diet as constant as possible, researchers can test whether differences carried by the transplanted community are capable of moving a phenotype.

The twin design sharpened the question. Sisters who shared much of their genetic background and early environment differed in obesity status. Their microbial communities could be transferred into mice that shared the same mouse genetics. If the recipient animals diverged, the difference could not be explained simply by the human twins' DNA.

The obese-donor communities drove greater adiposity under the experimental conditions. That result was striking. The more revealing part came when mice carrying different communities were housed together.

Mice are coprophagic. They encounter one another's microbes through the environment. In some diet conditions, organisms associated with the lean donor invaded the obese-donor community and blunted the phenotype. Under other diets, that rescue did not occur.

The community was causal, but not sovereign.

Host, microbes, diet, and ecological invasion interacted. The result resisted the two easiest headlines: *obesity is genetic* and *obesity is caused by gut bacteria*. Both statements were too complete.

The experiment instead offered a systems sentence: aspects of metabolic phenotype can be transmitted by a microbial community, and the expression of that capacity depends on ecological and dietary context.

That is the kind of sentence drug discovery finds difficult because it does not identify one permanent target. It identifies a state with several control points.

Seventy trillion cells, one metabolism

Anatomy is drawn with borders. Liver is separated from intestine, brain from blood, skin from environment. A cell diagram encloses cytoplasm within a membrane. A microbiome diagram places bacteria inside a tube labeled *gut* and draws arrows toward distant organs.

The borders are real. They are also busy.

Nutrients cross. Waste leaves. Immune cells sample. Hormones circulate. Bile exits the liver, enters the intestine, is transformed, and returns. Microbial products touch mucus and epithelium. Drugs encounter bacterial enzymes before or after host metabolism. Cells die and become raw material. Molecules are conjugated, deconjugated, oxidized, reduced, cleaved, and rebuilt.

A reference adult contains roughly thirty trillion human cells and a comparable number of bacterial cells, depending on body size, bowel contents, age, health, and methodology. The old textbook claim that bacteria outnumber human cells ten to one was an overestimate. Near parity is still extraordinary.

The body is not an organism plus a rounding error. It is a composite ecology.

Jeremy Nicholson, Elaine Holmes, and colleagues used metabolic profiling to describe humans as *superorganisms*, emphasizing combined host-microbial metabolism. The term can mislead if it implies harmony or equal interests. Human cells provide the host's architecture, circulation, nervous system, and reproduction. Microbial communities can support health, exploit the host, or shift between those roles.

The useful part of the metaphor is chemical authorship. Many molecules in a human sample cannot be assigned cleanly to one kingdom. Their existence depends on sequential transformations across host, microbe, food, medicine, and environment.

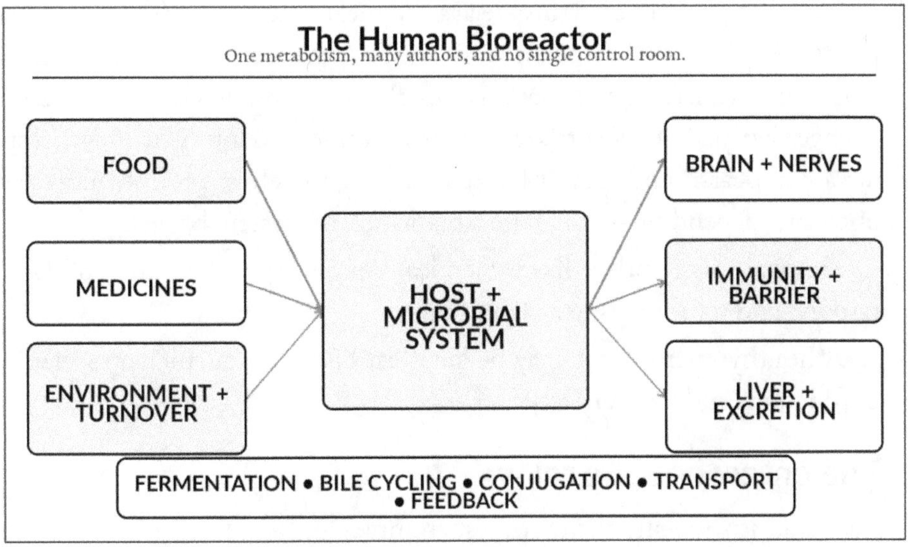

Figure 8. The human bioreactor: one metabolism, many authors, no single control room.

No central controller

An engineered bioreactor has sensors, valves, pumps, alarms, software, and an operator. The human system contains analogs of all of them, but no single controller sees the complete chemical state.

Control is distributed.

Local pH changes which microbes grow and which enzymes work. Oxygen falls steeply from tissue toward lumen, creating distinct niches across fractions of a millimeter. Mucus provides structure, attachment sites, and nutrients. Bile acids act as detergents, antimicrobial pressures, and signals. Microbes consume one another's waste. Host cells respond to microbial metabolites by changing barrier proteins, immune tone, and secretions. Those host changes alter the habitat, which changes microbial output.

The signal is also the condition. The output becomes the next input.

This recursive architecture is why reductionist experiments are both necessary and incomplete. Isolating a receptor can establish direct binding. It cannot show how binding changes the habitat that determines whether the ligand will exist tomorrow.

The system also contains delays. A meal changes substrates within hours. An immune response can reshape the community over days. Antibiotics can remove functions rapidly and allow new stable states to emerge. Early-life exposures may alter immune development long before disease appears. The same perturbation can therefore produce different outcomes depending on the state from which the system begins.

Homeostasis is not a fixed chemical composition. It is the ability to respond and return within a viable range.

A healthy microbiome may be less a list of organisms than a repertoire of functions and recovery trajectories.

The enterohepatic relay

Bile acids provide one of the clearest examples.

The liver synthesizes primary bile acids from cholesterol and commonly conjugates them before secretion into bile. After a meal, bile enters the intestine and helps solubilize fats. Most bile acids are later reabsorbed and returned to the liver through portal circulation.

Microbes intervene in the loop.

Bacterial enzymes deconjugate, oxidize, epimerize, and remove functional groups, generating a chemically diverse pool of secondary bile acids and related metabolites. These compounds differ in detergent properties, toxicity, antimicrobial effects, and receptor activity. They signal through pathways including FXR and TGR5, influencing metabolism, inflammation, and future bile-acid synthesis.

The liver provides starting material. Microbes rewrite it. The host reads the edited version and changes the next cycle.

There is no useful way to describe this as purely human or purely microbial metabolism. It is a circuit.

The aromatic-amino-acid relay

Tryptophan, phenylalanine, and tyrosine enter similar networks.

Tryptophan can be incorporated into protein, used in human serotonin and kynurenine pathways, or transformed by microbes into indole and related compounds. Specific bacterial routes generate circulating metabolites such as indolepropionic acid and other aromatic products. These molecules can influence receptors, barrier function, immune responses, and redox chemistry.

Phenylalanine and tyrosine feed microbial pathways that produce additional aromatic acids. In one influential study, Dylan Dodd and colleagues mapped a gut bacterial route that metabolizes aromatic amino acids into nine circulating metabolites. The work connected genes, enzymes, substrates, products, and host exposure rather than inferring function from a species list.

The lesson is not that every microbial product is beneficial. Some become harmful when kidney clearance is impaired or communities shift. The lesson is that food is chemically unfinished when it enters the mouth.

A Recent Signal from the Hidden Layer

A 2025 study offers a useful neighboring proof of principle for the argument in this book.

Researchers identified two approximately ten-kilodalton polypeptides, RORDEP1 and RORDEP2, produced by specific strains of the common gut bacterium Ruminococcus torques. The team detected the peptides in culture supernatants and in human plasma using targeted proteomics. In population datasets, the abundance of strains carrying the relevant precursor gene was inversely associated with body mass index and adiposity.

The investigators then moved up the causal ladder. RORDEP-expressing bacterial strains and engineered bacteria improved several metabolic outcomes in mouse models. Recombinant RORDEP1 altered gut-hormone responses and glucose regulation in rodents, and intestinal delivery changed liver pathways involved in glucose production and insulin signaling.

The limits matter. These are polypeptides rather than the small-molecule class at the center of this book. The human observations were associations and measurements of circulating material, not a trial showing weight loss or metabolic benefit in people. The functional interventions were preclinical, and the direct receptor mechanism remains incomplete.

What the study demonstrates is narrower and still important: a microbial product that standard host-centered diagrams did not include can be present in human blood, vary with microbial strain biology, and produce coordinated metabolic effects in animals. The molecule was not supernatural. It was simply outside the usual accounting frame until the right genomic, proteomic, culture, and intervention tools were connected.

This is how the ghost layer should become knowledge: not through one sweeping claim, but through a chain from hidden capacity to physical product, physical product to exposure, and exposure to controlled physiology.

The same meal enters two different chemical worlds

A meal is a chemical library, not a nutrition label. Proteins, branched carbohydrates, lipids, polyphenols, minerals, processing products, and microbial material enter a host-microbial system whose access tools differ from person to person.

Human enzymes reach some pages. Microbial enzymes reach others. Cooking, grinding, fermentation, transit time, and community structure determine which pages are opened and what is written in the margins. The same food can therefore enter two people and emerge as two different molecular experiences.

Nutrition science often begins with the composition of the meal. The bioreactor perspective begins with what the meal becomes.

Two people can eat the same beans, berries, meat, bread, or medication and generate different downstream chemistry. The difference can arise from host enzymes, transit time, bile-acid pools, microbial strains, phages, prior meals, circadian state, kidney function, or the physical structure of the food.

A fiber listed as one nutrient on a label can contain a family of linkages that different microbial communities access with different efficiency. One person may convert more of it into butyrate. Another may route carbon toward acetate, lactate, gas, or biomass. Cross-feeding can turn one organism's product into another organism's substrate. The endpoint is not written on the package.

Protein creates a similar branching problem. In the small intestine, host proteases release peptides and amino acids. Material that reaches the colon encounters microbial proteases and fermentative pathways. The resulting products can include indoles, phenols, amines, sulfur compounds, and branched-chain fatty acids. Their effects depend on dose, clearance, barrier state, and the larger molecular mixture.

Food processing changes the experiment before digestion begins. Grinding increases accessible surface. Heating unfolds proteins and creates new reaction products. Fermentation performs microbial transformations outside the body. Emulsification changes interfaces. An intact plant cell wall delivers substrates differently from a purified powder with the same nominal carbohydrate.

This does not make every dietary response mysterious. Energy balance remains real. Essential nutrients remain essential. Clinical evidence remains the standard. It means that a label describing inputs cannot fully predict outputs in a living ecosystem.

The future of personalized nutrition will therefore need a functional readout. Species lists alone are unlikely to be enough. Two communities with different memberships can converge on similar metabolites, while similar taxonomic profiles can diverge because strains, genes, expression, phages, and substrate history differ.

The most useful question may be dynamic: *How does this person's system transform a standardized challenge, and how quickly does it return to a resilient state?*

That is closer to a glucose-tolerance test than a personality quiz for the microbiome. It measures response rather than declaring identity.

Medicines enter the same reactor

Oral pharmaceuticals are not exempt from microbial chemistry.

The liver receives enormous attention in drug metabolism because cytochrome P450 enzymes and conjugation pathways can determine exposure, interactions, and toxicity. The intestine contains an additional metabolic organ whose enzymes vary across people.

Some strains of *Eggerthella lenta* can inactivate digoxin under particular conditions. Bacterial beta-glucuronidases can remove host-added glucuronide groups from drug metabolites excreted into the intestine. In the case of irinotecan, reactivation of a toxic metabolite in the gut contributes to gastrointestinal injury; inhibiting selected bacterial enzymes reduced that toxicity in experimental models.

These examples reveal a hidden source of patient variability. Two people can swallow the same tablet and generate different chemical exposures before or after the drug reaches the intended target.

A full ADME model should therefore include microbial transformation where relevant. Absorption, distribution, metabolism, excretion, and toxicity are properties of the host-microbial system, not the liver alone.

The implication cuts both ways. Microbial enzymes can be liabilities to avoid. They can also be used intentionally to activate prodrugs or local therapies. A molecule designed to remain inert until it reaches a community-specific enzyme could create targeted release. The challenge is variability: the activating function may be absent, unstable, or shared with unintended organisms.

Drug development needs to measure the capability rather than assume it.

Cell death is part of the feedstock

The human body continuously replaces cells. Intestinal epithelium turns over. Immune cells die after activation. Red blood cells are dismantled. Damaged components are recycled through autophagy. Microbial cells grow, divide, lyse, and become nutrients for neighbors.

The word *waste* hides this circulation.

A dying cell releases or exposes lipids, nucleotides, peptides, sugars, heme, cofactors, and modified proteins. Host phagocytes process some material. Microbial enzymes process other material. Molecules move into bile, urine, breath, stool, or new biomass. Some fragments signal before they are cleared.

The factory consumes itself to remain alive.

This continuous turnover expands chemical possibility beyond de novo biosynthesis. A molecule need not be made from scratch. It can be clipped from a larger structure, modified, conjugated, folded, complexed, and reused. Proteolysis is not only destruction; it can reveal bioactive peptides. Lipid remodeling creates signaling mediators. Oxidation can deactivate a compound or create a reactive signal. Microbial enzymes can reverse host conjugation and reopen a molecule for further transformation.

The final product can carry a chain of authors.

The cryptic-assembly hypothesis

Think of cryptic assembly as a relay race in which no runner carries the final baton from start to finish.

A host enzyme may release a fragment. A microbe may trim it. A neighboring organism may add a lipid or remove a sugar. Local oxygen, metals, pH, or redox state may change the product again. The final active entity can be real even though no single genome contains a complete recipe written from beginning to end.

The conservative form of this idea is already established in bile-acid, aromatic-amino-acid, polyphenol, and drug metabolism. The stronger hypothesis is that distributed processing also creates previously unrecognized constrained peptides, cyclic products, hybrid lipids, complexes, or transient signaling states.

That stronger claim must be tested. A possible sequence looks like this:

1. A food or host protein is cleaved into a short peptide.
2. A microbial or host enzyme trims, ligates, acylates, oxidizes, or otherwise modifies the fragment.
3. A second organism or tissue performs another transformation.
4. Local conditions favor ring closure, metal coordination, disulfide formation, or a stable conformation.
5. The resulting molecule acts locally and is rapidly degraded.

Pieces of this sequence occur throughout biology. The stronger claim - that such distributed assembly produces a large, underrecognized class of human signaling molecules - remains unproven.

Why might standard experiments miss it?

Pure culture removes neighboring organisms. Defined media remove dietary precursors. Bulk tissue extraction destroys spatial gradients. Delayed processing allows reactive intermediates to vanish. Genomic prediction searches for one organism's biosynthetic machinery. Targeted metabolomics measures known compounds. Standard peptide analysis may assume linear products.

Each method is rational. Together they can erase the assembly line.

The hypothesis makes testable predictions. Some molecular features should appear only when intact communities encounter specific substrates under near-native conditions. Stable-isotope labels should trace atoms from food, microbe, and host into the same product. Removing one community member should eliminate the product or cause an intermediate to accumulate. Synthesizing the candidate should reproduce the spectrum and phenotype.

The question is not whether the story sounds plausible.

The question is whether the atoms can be followed.

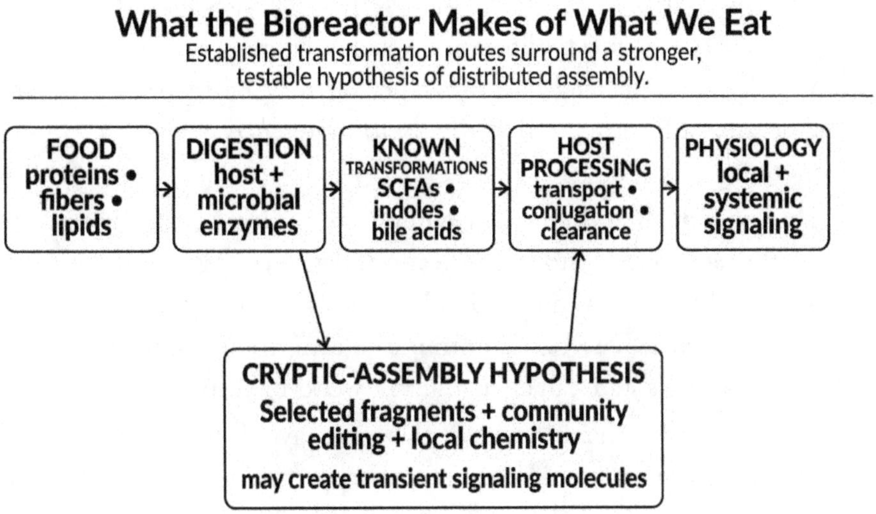

Figure 9. What the bioreactor makes of what we eat. Solid paths are established transformations; the dashed branch is the stronger cryptic-assembly hypothesis.

The scale of flux

A static inventory asks how many molecules are present. A living-system inventory must also ask how quickly they move.

A metabolite can remain at low abundance because it is produced and consumed rapidly. At a receptor-rich boundary, that high flux can carry more biological information than a stable compound present at much higher concentration. Blood sampling favors molecules that survive circulation. Stool favors material that reaches the end of the gut. Urine favors what the kidney clears.

Imagine trying to understand a city by analyzing its landfill. The landfill contains real evidence. It does not capture every purchase, traffic signal, text message, or decision that occurred during the day.

This is why time-resolved challenge studies can be more informative than fasting snapshots. A standardized meal, medication, exercise bout, infection, or sleep disruption perturbs the reactor. The trajectory - which molecules rise, which intermediates appear, how quickly the system returns - can reveal functional capacity.

The health marker may be resilience rather than baseline.

A person whose metabolite panel appears ordinary at rest may respond abnormally to a challenge. Another person may begin from a different baseline and still return efficiently. Static reference ranges can miss the distinction.

The individualized bioreactor complicates universal biomarkers. Diet, geography, age, birth history, medication, pets, occupation, and infection shape microbial communities. Similar species can carry different strains, phages, and genes. Different communities can produce similar functions. The same person's state can change across weeks or years.

Personalized medicine should therefore avoid the fantasy of one universally healthy microbiome. The more practical objective is to identify chemical functions and trajectories associated with resilience in a defined host and indication.

Disease as a chemical state

The twins-and-mice experiment changed the meaning of microbial causality. The phenotype did not travel as one pathogen or one toxin. It traveled as a community capable of producing a different metabolic state in a host.

This is a more useful way to think about many chronic diseases.

Dysbiosis is often described as a bad mixture of microbes. Composition is only one dimension. A community can be taxonomically altered yet functionally resilient, or taxonomically familiar yet chemically dangerous. Disease may reflect changes in substrate flow, barrier geography, phage pressure, cross-feeding, host clearance, or the timing of chemical responses.

The clinically useful marker may therefore be a pattern - a set of molecules, ratios, and trajectories - rather than one organism.

A snapshot can also confuse cause and consequence. A diseased tissue changes oxygen, nutrients, immune pressure, medications, and movement. Those changes alter the microbiome and metabolome. A molecule associated with disease may be a driver, a defense, a compensation, or debris.

Time matters. The most informative signal may appear before symptoms, during transition, or after intervention. The bioreactor is a process. We need recordings, not photographs.

The gut-brain axis without magic

Few phrases have traveled farther from their evidence than *gut-brain axis*.

The phrase is valid. The gut and brain communicate through nerves, hormones, immune signals, metabolites, vascular pathways, behavior, and shared physiology. Microbial products can influence cells that release peptides, alter immune tone, affect barrier function, or enter circulation. The vagus nerve carries information in both directions. Stress changes motility and secretion, which changes microbial habitat. Sleep and diet alter the system again.

The mistake is imagining a simple pipe from a bacterium to a thought.

Serotonin is a common example. Much of the body's serotonin is produced in the gut, largely by host enterochromaffin cells, and microbial signals can influence that production. Peripheral serotonin does not simply cross the blood-brain barrier and become mood. Its local and systemic roles include motility, platelets, vascular effects, and metabolic signaling. Brain serotonin is regulated within the central nervous system.

A microbial metabolite can still affect the brain without entering it. It may act on an intestinal cell that releases a hormone, on an immune cell that changes cytokine tone, on a sensory neuron, or on the integrity of a vascular or epithelial barrier. The causal chain can involve several relays.

This is precisely why the ghost-genome argument matters. A target-only assay can miss the first meaningful step because the clinically relevant outcome is separated from the initial molecular interaction by tissue geography and network response.

It is also why claims require discipline.

An association between a microbial feature and depression, autism, Parkinson's disease, Alzheimer's disease, or anxiety is not a diagnosis or a mechanism. Mouse behavior is not a human psychiatric outcome. A probiotic that changes a survey score in one small study is not proof that the microbiome controls mood.

A credible program must establish a chain:

- the molecule or pathway is present and measurable;
- the relevant exposure reaches a defined cell or compartment;
- perturbing the molecule changes a proximal biological event;
- the proximal event changes a disease-relevant phenotype;
- the result reproduces across models and, eventually, carefully selected patients.

The gut-brain axis becomes more convincing when it is described less magically.

It is not one axis. It is a network of chemical and electrical routes whose importance changes by disease, person, and time.

Several ways to change the same output

If a missing or harmful microbial product contributes causally to disease, medicine can intervene at several levels.

The molecule can be supplied directly. A precursor can be added. The organism that makes the product can be introduced or encouraged. A competitor that consumes the molecule can be inhibited. Diet, pH, transit, oxygen, or bile flow can be altered to favor the pathway. A host clearance mechanism can be adjusted. A synthetic analog can reproduce the desired activity without depending on the community.

Each strategy has a different risk.

Direct delivery offers dose control but may miss the timing and location of native production. A live organism can adapt and produce dynamically but may fail to engraft or may behave differently in another host. A dietary precursor is accessible but can feed several competing pathways. An enzyme inhibitor can be precise but requires a validated causal map. A synthetic analog may be controllable but lose the ecological pattern that made the natural molecule useful.

The mature therapy may combine levels.

This is why the chemical output is a powerful organizing point. Taxonomy tells us which organisms might be manipulated. Chemistry tells us what state the intervention is trying to create.

Personalization means preserving causality

The word *personalized* is often used when a product has merely become more complicated.

A true personalized bioreactor intervention would not begin by assigning every patient a unique supplement list. It would identify a causal bottleneck that differs across a meaningful subgroup.

One person may lack a microbial conversion needed to activate a dietary precursor. Another may produce the desired metabolite but clear it rapidly. A third may carry the pathway yet fail to express it under the relevant diet. A fourth may generate a harmful competing product. The intervention could therefore differ at several levels:

- provide the final molecule;
- provide a precursor only to people who can convert it;
- add or encourage a producer;
- inhibit a competing enzyme;
- change diet or local conditions;
- bypass the ecology with a synthetic analog;
- select a different therapy because the pathway is not relevant in that patient.

Personalization is useful only when the measurement changes the decision.

The twin experiment illustrates the principle. The community effect depended on diet and ecological invasion. A treatment based only on the presence of one species would have missed the interaction. A treatment based only on the host phenotype would have missed the transmissible capacity.

The most valuable biomarker may be neither taxonomic nor purely clinical. It may be a transformation signature: the measured ability of the system to move a defined substrate through a defined molecular route.

That is a harder assay to build. It is also closer to mechanism.

The rescue hidden in the cage

The most hopeful part of the twins experiment was not that an obesity-associated phenotype could be transmitted. It was that the outcome could be modified.

A lean-associated community entered the ecological space of the other mice and changed the trajectory - but only under dietary conditions that supported invasion. The intervention was not a magic organism acting independently of context. It was ecology.

That observation foreshadows a different kind of medicine. Instead of asking only how to block one target after disease is established, we may learn to alter the chemical state that makes the disease resilient. The intervention could be a molecule, a precursor, a microbial strain, a phage, an enzyme inhibitor, a dietary structure, or a combination. In some cases, the safest strategy may be to deliver the chemical instruction directly rather than asking a living organism to establish itself.

The body is modifiable because the bioreactor is dynamic.

It is also dangerous to manipulate because the network can move in directions we did not intend.

The next story enters the mucus layer, where viruses that do not infect human cells can nevertheless become part of a protective boundary. It is a story about function emerging from physical arrangement - and about chemical events that appear only when biology is observed in place.

CHAPTER 6

The Viruses in the Mucus

The Immune Layer the Host Did Not Encode

STORY 6

THE VIRUSES IN THE MUCUS

The Immune Layer the Host Did Not Encode

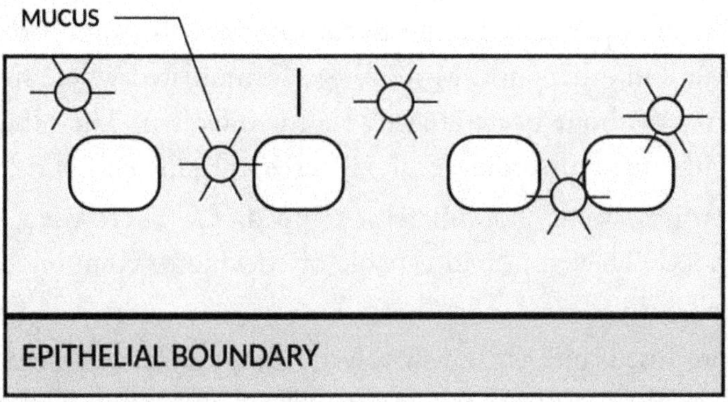

THE TURNING POINT

Function emerged from an arrangement shared across kingdoms.

Jeremy Barr was studying two things that did not seem to belong in the same sentence: bacteriophages and mucus.

Phages are viruses that infect bacteria. Mucus is the hydrated, glycoprotein-rich material coating exposed tissues in the mouth, lungs, gut, and other surfaces. Phages are abundant wherever bacteria live, so finding them in mucus was not surprising. Mucus contains bacteria. Bacteria attract their viruses.

The surprise was that phages appeared enriched there.

Barr, working in Forest Rohwer's laboratory, and his colleagues found higher phage-to-bacteria ratios in mucus than in surrounding environments across diverse animals. A simple explanation came first: perhaps the viruses were replicating more because mucus contained more bacterial hosts.

The team removed the hosts from the question.

They applied phage to sterile tissue-culture cells, some producing a mucus layer and others not. More phage accumulated where mucus was present, even without bacteria available for infection. They then coated surfaces with mucus macromolecules. Phages still adhered.

The viruses were not merely trapped in a crowded bacterial neighborhood. They appeared capable of sticking to components of the mucus itself.

The proposed mechanism involved immunoglobulin-like domains displayed on phage capsids interacting with mucin glycans. The result was a striking ecological arrangement. The animal produced the mucus. Phages accumulated within it. Bacteria moving toward the epithelial surface encountered a concentrated field of their viral predators.

In experimental systems, the mucus-associated phages helped protect underlying cells from bacterial attack.

Barr and colleagues called the model **Bacteriophage Adherence to Mucus**, or **BAM**.

The host genome did not encode the complete protective layer. The function emerged from mucus chemistry, phage structure, bacterial movement, and spatial concentration.

No single participant authored the system.

That is the ghost genome in miniature.

A crowded border

A mucosal surface is one of the busiest borders in biology.

The intestine, airway, oral cavity, reproductive tract, and other exposed tissues cannot build an impermeable wall. They must exchange gases, nutrients, water, signals, and cells while limiting invasion. Mucus helps solve the contradiction.

It is a hydrated mesh of large glycoproteins, salts, lipids, antibodies, antimicrobial factors, cellular material, and microbial products. Its structure changes with location and disease. In parts of the colon, the inner layer normally restricts bacterial contact with the epithelium while the outer layer supports dense microbial life. In the airway, cilia move the mucus and its trapped contents. At every site, the material is being secreted, degraded, modified, and displaced.

To a bacterium, mucus is habitat and obstacle. To the host, it is barrier and sampling zone. To a phage, it can be a place of retention.

The insight behind the bacteriophage-adherence-to-mucus model was spatial rather than taxonomic. A phage did not need to be encoded by the host or permanently integrated into a host cell to contribute to defense. If capsid features interacted with mucin glycans, the virus could spend more time in the layer through which bacterial hosts moved.

The probability of encounter changed.

That is a small sentence with large implications. Biological function can arise by concentrating participants in the same place. Neither a phage genome nor a mucin gene alone specifies the complete outcome. The host produces the matrix. The phage carries the binding surface. The bacterium supplies the replicative host. Flow and diffusion set the timing.

The protective effect lives in the arrangement.

This kind of function is difficult to discover in homogenized samples. Grind the tissue, dilute the mucus, and average the phages across a tube, and the spatial advantage disappears from the measurement even if every component remains present.

The ghost genome includes these relational states. The hidden entity is sometimes not a novel molecule. It is a known set of molecules placed into a functional geometry.

Mucus is not slime around the real tissue

Mucus is often treated as a nuisance in experiments - a viscous material to wash away before reaching the cells beneath it.

In the body, it is an active interface.

Mucins are large, heavily glycosylated proteins that absorb water and create a structured gel. The layer traps particles, shapes diffusion, provides binding sites, and creates a habitat for microorganisms. Its thickness, composition, and turnover vary by tissue and physiological state. In the intestine, distinct mucus regions can separate dense microbial populations from the epithelial surface while still allowing chemical exchange.

The glycans decorating mucins are not inert ornament. They can serve as adhesion sites and nutrient sources. Host enzymes, microbial glycosidases, antibodies, antimicrobial peptides, metabolites, and phages all operate within the same material.

A molecule or virus entering mucus experiences a different physical world from bulk liquid. Movement slows. Local concentration changes. Binding becomes possible. Enzymes and receptors encounter one another in confined space.

This is why removing mucus can simplify an assay and erase the mechanism at the same time.

Barr's experiments followed the logic of alternative explanations. If phages were enriched only because bacteria multiplied in mucus, sterile mucus-producing cells should not accumulate them. They did. If the gel merely slowed diffusion through mass action, purified mucus components attached to a surface should not show selective adhesion. They did. The observations moved the explanation from coincidence toward interaction.

The proposed capsid-glycan mechanism was not the final word on every phage or mucosal surface. It made the spatial phenomenon experimentally tractable.

Function can live in an arrangement

Biology is usually cataloged as things: genes, species, proteins, cells, molecules. The BAM model draws attention to an additional unit - arrangement.

The same phage in a test tube, the intestinal lumen, and a mucus layer does not have the same ecological meaning. In mucus, adhesion changes encounter rates. The gel slows motion. Bacteria supply hosts. The epithelium supplies a vulnerable boundary. A function appears because the components occupy a particular geometry.

There may be no gene for *phage shield*.

There are genes for mucins, phage capsid proteins, bacterial receptors, and the machinery that builds each participant. The protective phenomenon arises from their physical relationship.

This distinction matters for chemistry. A molecule's concentration in blood can be low while its concentration at a surface is high. A ligand can be irrelevant in bulk fluid and potent inside a membrane, biofilm, mucus layer, or protein complex. A transient product can alter a local receptor before diffusion and clearance make it nearly invisible systemically.

The conventional sample often destroys the arrangement before measurement. Homogenize a tissue and mucus, epithelium, microbial cells, and immune compartments become one average. The average may contain every component and still conceal the function.

Spatial biology is not decorative context around the molecule.

Location can be part of the mechanism.

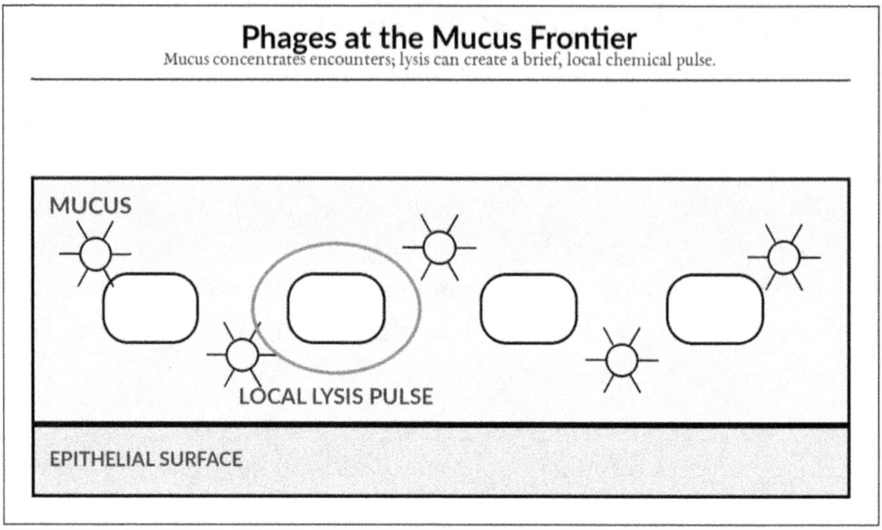

Figure 10. Phages at the mucus frontier: adhesion, bacterial encounters, lysis, and local chemical release.

The viral world inside the microbial world

The number 10^31 belongs to Earth, not to one human intestine. It is a global estimate for virus-like particles across oceans, soils, sediments, and living systems. The human virome is smaller and still immense.

In the gut, most viral sequences detected in healthy populations are associated with bacteriophages. Large catalogs contain well over one hundred thousand nonredundant gut phage genomes, with much diversity unclassified. The numbers vary with sampling and method, and a sequence catalog does not tell us which phages are active.

Phages do not merely add another list to the microbiome.

A lytic phage infects a bacterium, replicates, and breaks the cell open. A temperate phage can integrate into the bacterial genome or persist in another state, changing bacterial phenotype without immediate lysis. Environmental conditions can shift the relationship.

Through these interactions, phages can reduce bacterial populations, release cellular contents, move genes through transduction, alter virulence and metabolism, impose selection for resistance, influence biofilms, and restructure community networks.

Calling them *conductors* is a metaphor, not a claim that phages direct the microbiome toward host health. A conductor follows a score. A phage follows replication. Its effects can still change which chemical voices become loud, quiet, or absent.

The long life of a temperate phage

Not every phage kills immediately.

A temperate phage can enter a bacterium and persist, often by integrating genetic material into the bacterial chromosome or maintaining another stable form. The viral genome becomes a prophage, copied as the bacterium divides. Under certain stresses or regulatory changes, it can return to active replication and lysis.

This lifestyle blurs the categories of predator and passenger.

A prophage can alter its host's phenotype. It can contribute genes, regulatory elements, toxins, metabolic functions, or immunity against related phages. The bacterium that enters the community is no longer chemically identical to the bacterium before infection. Its competitors and host interactions may change even while no viral particles are visibly bursting from cells.

For the ghost-genome thesis, the important point is that microbial chemical capacity is historically contingent.

Two bacterial strains that look closely related by ordinary taxonomy may carry different prophages. The difference can alter gene expression, stress response, virulence, surface features, or susceptibility to later infection. A metagenomic snapshot may detect the viral sequence without revealing whether it is active, silent, defective, or controlling a nearby bacterial gene.

Antibiotics, inflammation, oxidative stress, or other disturbances can induce some prophages. The resulting lysis can restructure the community and release bacterial contents. A treatment aimed at one organism can therefore create a chemical and ecological second-order effect.

This is one reason microbiome interventions cannot be evaluated only by the abundance of named bacteria. The same apparent community can contain different viral histories and different probabilities of transition.

The virome is not background decoration around the microbiome.

It is part of the memory by which the microbiome became its present state.

Predator, passenger, or gene cassette

Phages occupy several ecological roles because their life cycles differ.

A strictly lytic phage enters a susceptible bacterium, redirects cellular machinery, assembles progeny, and ruptures the host. The immediate population effect can be dramatic.

A temperate phage can integrate into the bacterial chromosome as a prophage or persist in another stable form. The bacterium survives and replicates with viral DNA inside it. The prophage can alter gene expression, provide new traits, exclude competing phages, or later be induced into a lytic cycle by stress.

Some phages carry accessory genes that change bacterial virulence, metabolism, or ecological fitness. Transduction can move bacterial DNA between cells. The transfer is not a deliberate software update and often fails to establish, but over evolutionary time phages participate in the circulation of microbial functions.

This means the chemical repertoire of a community is not fixed. A pathway can be gained, lost, silenced, or activated. A strain can become more toxic, more competitive, or metabolically different because of a virus.

The personal virome reflects this dynamism. Individuals often carry distinctive, partly stable phage communities. Antibiotics, diet, inflammation, infection, and other disturbances can change them. A sequence catalog can identify viral genomes without revealing which are replicating, integrated, or influencing chemistry at that moment.

Once again, identity is not function.

Lysis is a chemical event

When a phage lyses a bacterial cell, the event is usually described in biological terms: the host dies, new viruses are released, the population changes.

It is also an abrupt chemical release.

A bacterial cell contains nucleic acids, proteins, peptides, metabolites, ions, membrane fragments, and specialized products at concentrations far higher than the surrounding fluid. Lysis creates a local pulse. Neighboring microbes may consume the material. Host receptors may encounter cell-wall components. Enzymes may continue acting briefly after release. Reactive compounds may be created or destroyed.

It is tempting to call this a delivery system. That would imply purpose. The release is a consequence of viral replication, not an intervention designed for the animal host.

The consequence may still matter.

The stronger ghost-genome hypothesis predicts that some transient chemical signals arise specifically during community events such as phage infection, bacterial competition, cell death, or biofilm disruption. These signals would not appear as steady secretion from a pure culture. A bulk sample might average them away. A delayed sample might miss them completely.

Testing the hypothesis requires time-resolved, spatially resolved measurement during controlled phage-bacteria interactions. It also requires distinguishing meaningful signals from the nonspecific chemistry of cellular damage.

A local pulse can be biologically powerful and analytically easy to lose.

How to see a microburst

A lysis-generated chemical pulse presents a difficult measurement problem.

The event is local. It may occur in a biofilm, mucus pocket, crypt, or inflamed region. The released molecules can be consumed, degraded, oxidized, or diluted within seconds. By the time a stool or blood sample is collected, the original pulse may be represented only by downstream products.

A useful experiment would synchronize or monitor a defined phage-bacteria interaction in a spatial model. Samples would be taken before infection, during replication, at lysis, and through recovery. High-resolution metabolomics would be paired with bacterial counts, phage counts, transcription, imaging, and isotope labels. A parallel system lacking the phage receptor or lysis function could distinguish viral effects from ordinary bacterial stress.

The strongest candidate signals would be features that appear reproducibly at the event, depend on the interaction, reach plausible concentrations near host cells, and produce a phenotype when synthesized or reintroduced.

This is expensive, but the design illustrates an important principle: the sampling schedule must be built around the biological event rather than administrative convenience.

Weekly stool collection cannot answer every question about chemistry that exists for thirty seconds in mucus.

Ghost Activity Can Precede Identity

A bacterial cell is a concentrated chemical compartment, and a phage-driven rupture can create an effect before science knows which released molecule caused it.

That order is common in natural-products discovery. An extract, fraction, or timed sample changes a phenotype. The activity survives repetition but disappears when the material is separated incorrectly, delayed, oxidized, or diluted. At that stage the activity is evidence, not identity.

Activity-guided fractionation turns the effect into a trail. The sample is divided, each fraction is retested, and the active neighborhood is divided again. Spectral features that repeatedly travel with the phenotype become suspects. Recombining fractions can reveal synergy. Time-resolved samples can show whether the effect belongs to simple release, a newly formed product, or a later degradation state.

The discipline is to keep the language exact. A reproducible fraction can contain a real ghost activity. It does not yet justify a named molecule, a target, or an evolutionary purpose. Those conclusions arrive only after purification, structural confirmation, synthesis or another authentic standard, exposure measurement, and causal controls.

This sequence is especially important for short-lived events such as lysis. The biological event tells us when to look. The activity tells us which material to follow. Chemistry tells us what was there. Intervention tells us whether it mattered.

The activity is not the answer. It is the first reliable footprint.

The three layers of the ghost orchestra

A mass spectrometer does not hear music. It measures mass, charge, time, and fragments. The orchestra is our interpretation.

In an untargeted metabolomics experiment, a complex sample produces thousands of features. Some match authentic standards. Some receive tentative formulas or structural classes. Some are isotopes, adducts, fragments, contaminants, or artifacts. Many remain unknown.

Unknown does not mean important.

Unknown also does not mean unimportant.

It means the instrument detected something and the reference system could not confidently name it.

To keep ignorance from becoming mythology, the ghost orchestra can be divided into three layers.

Known molecules with underappreciated roles. Short-chain fatty acids, secondary bile acids, indoles, aromatic metabolites, microbial lipids, cell-wall fragments, and many others are structurally known. Their tissue-specific, dose-specific, temporal, and combinatorial functions remain incomplete.

Reproducible features with incomplete identities. These are recurring spectra or molecular families without fully confirmed structures. Some will prove to be variants of known compounds. Some will be artifacts. Some will be new chemistry.

Hypothesized transient or context-dependent entities. These may not survive conventional sampling. They include unstable intermediates, local complexes, cryptically assembled products, metal-bound states, and brief molecular pulses. This layer is the most provocative and the easiest to overclaim.

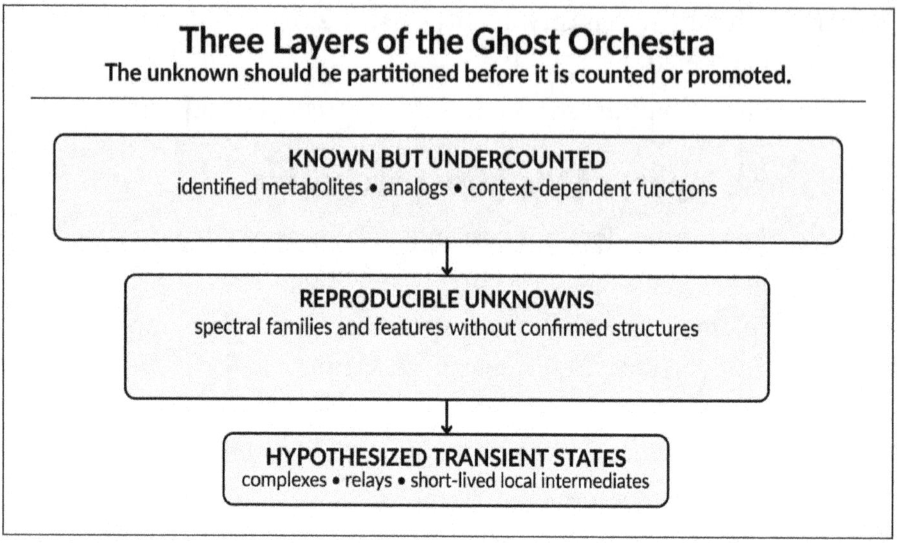

Figure 11. The ghost orchestra: known but undercounted chemistry, reproducible unknowns, and hypothesized transient states.

The BAM story touches all three layers. The phages and mucins are known. The full range of molecular effects created by phage adhesion, infection, lysis, and bacterial response is only partly mapped. Some consequences may exist only at the boundary and for seconds or minutes.

Identity is not function

Human gut viromes are highly individualized. Many phages remain stable within a person while diet, disease, antibiotics, and other disturbances can alter the community. A phage sequence can suggest lifestyle or host range. It cannot tell us by itself whether the phage is actively replicating, integrated, inducing bacterial stress, or changing a metabolite relevant to the host.

This is the recurring limit of omics.

Identity describes who or what could participate. Function requires state, location, substrate, and time.

The same distinction applies to bacteria. Two people may carry similar species but different strains, phages, mobile genes, expression states, and chemical outputs. Conversely, taxonomically different communities can perform similar functions.

A clinically useful microbiome measurement should therefore move beyond the question *Who is there?* toward *What is happening, where, and with what chemical consequence?*

The promise and hazard of precision predators

Antibiotic resistance has revived interest in phage therapy because a virus can be exquisitely specific for a bacterial strain. Specificity can preserve much of the surrounding community and allow treatment of bacteria resistant to standard drugs.

The same specificity creates operational problems. A phage may fail against a slightly different strain. The bacterium can evolve resistance. The patient's immune system can neutralize the virus. A cocktail may need to be updated. Regulatory and manufacturing systems must control identity, purity, potency, and genetic safety while the biological product itself can evolve.

Killing bacteria can also release inflammatory components or toxins. Temperate phages can move undesirable genes. Removing one organism can open ecological space for another.

A ghost-genome-informed phage program would therefore measure chemical consequences as well as bacterial burden. Does the intervention reduce a harmful metabolite? Does it release endotoxin or another inflammatory signal? Does a beneficial cross-feeding function disappear? Does the community return to a resilient state or settle into a worse one?

The therapy is a perturbation of an ecosystem, even when the immediate target is one strain.

This is not an argument against phages. It is an argument for treating precision at the level of identity as only one dimension of precision.

Therapeutic promise without ecological innocence

Phage therapy is regaining attention as antibiotic resistance grows. Phages can be highly specific and can attack bacteria resistant to conventional drugs. They can also evolve, miss strains, be neutralized, release inflammatory material, or alter communities in unintended ways.

Manufacturing and regulation are challenging, especially for personalized cocktails.

The ghost-genome perspective suggests another therapeutic dimension. A phage intervention might be evaluated not only by pathogen counts but by the chemical ecology it changes. Removing one strain could reduce a harmful metabolite, release a beneficial function from competition, or destabilize a community in ways that matter later.

That possibility is not routine medicine. It is a reason to measure more carefully.

The human bioreactor is not a symphony composed for our well-being. It is a negotiated ecology under continuous selection. A phage bloom can destabilize a beneficial community. Lysis can release toxins. Gene transfer can spread virulence. The same process that creates diversity can create disease.

Medicine's task is not to restore an imaginary pristine harmony. It is to identify states associated with resilience, establish which chemistry is causal, and intervene with enough precision to move the system without creating a worse alternative.

The virome as a patient-specific variable

Human gut viromes are highly individualized. Even when broad functional patterns recur, many viral sequences differ from person to person, and large fractions cannot be confidently assigned to known hosts or lifestyles.

That individuality can influence therapy.

A phage intended to remove a pathogenic bacterium must encounter a susceptible strain. The target can evolve resistance. The patient's immune system may clear the phage. Mucus binding can change distribution. Other members of the community may fill the vacated niche. Lysis can release inflammatory material. A temperate phage may transfer genes rather than simply kill.

These are not reasons to abandon phage therapy. They are reasons to treat a phage as a biological intervention rather than a narrow antibacterial reagent.

The same logic applies to drugs and diet. An antibiotic can induce or suppress phage dynamics. A dietary change can alter bacterial growth and therefore viral replication. Inflammation changes the habitat. The patient's virome becomes one variable in the response to an intervention that was not designed as antiviral.

Future trials may need to measure that variable, especially when the therapeutic effect depends on microbial community remodeling. A baseline virome could help explain why one patient's bacterial target collapses while another's rebounds. Longitudinal viral and metabolomic data could reveal whether a clinical response follows removal of a harmful function, release of a transient product, or ecological replacement.

This is an example of the book's larger argument. The hidden chemical language does not live only in molecules. It lives in events that determine when molecules appear.

A phage infection is one of those events.

A border made by three kingdoms

The mucus layer returns later in this book when we discuss geofenced medicines. It shows why local exposure can be more than a safety tactic. A therapy acting in mucus enters an ecosystem containing host polymers, immune molecules, bacteria, phages, food-derived material, and rapidly changing gradients.

A compound that never reaches high concentration in blood may still reshape the boundary where disease begins.

The BAM model also provides a warning for artificial intelligence. A database can contain the sequences of mucins, phages, and bacteria without containing the emergent function produced by their arrangement. Unless the model receives spatial and experimental context, the protective layer does not exist in its training data.

The function is real.

The representation is incomplete.

The next part of the book turns from what the old map missed to the instruments that can reach it. The first of those instruments began, for me, not with a mass spectrometer or an algorithm, but with pumps, fouling, water, and the realization that a lake-sized problem could not be solved in a vial.

PART III

Build the Telescope

Two stories about changing the experiment before asking nature for an answer.

• • •

CHAPTER 7

The Lake That Would Not Fit in a Petri Dish

Scale Becomes an Instrument

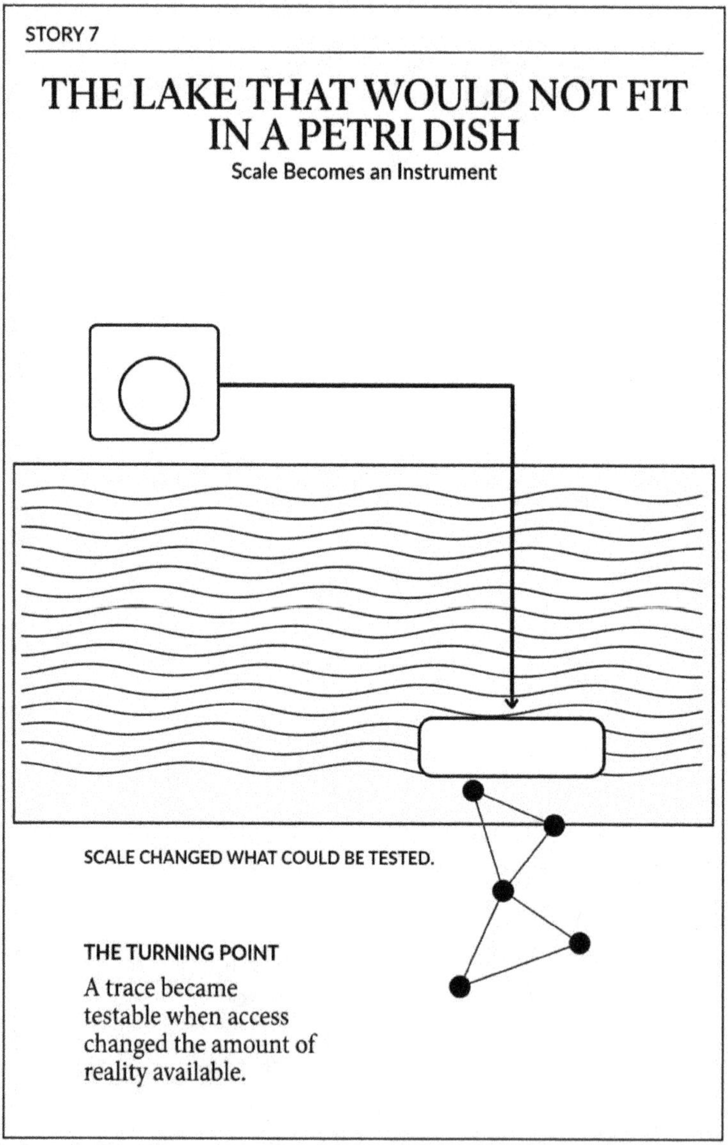

The microbial world first became real to me at a scale that did not fit on a laboratory bench.

A dense aquatic bloom looks simple from a distance. The surface changes color. Wind pushes material into streaks and coves. A sample jar fills with green or brown water. To most people, it is an environmental problem, a warning sign, or an unpleasant day at the lake.

To a natural-products chemist, the same bloom is a living factory.

The difficulty is that the factory has no loading dock.

The chemistry is dispersed across water, cells, membranes, particles, viruses, extracellular material, and dissolved organic matter. Some molecules are present at concentrations so low that a conventional sample contains only a trace. Some disappear when the community is removed from light or oxygen. Some are produced only when one organism is crowded by another. Some may be intermediates that never accumulate long enough to become a clean peak.

The standard solution is sensible: take a sample, bring it to the laboratory, isolate whatever will grow, and ask the survivors to reproduce their chemistry in a flask.

It is also an enormous filter.

The question that shaped our work was physical before it was computational:

What if we stopped asking the ecosystem to become a laboratory culture and built a process large enough to meet the ecosystem where it was?

That question led to pumps, separators, membranes, hoses, trucks, weather, preservation, extraction, fractionation, and a great deal of failure.

It also led back to the first instruments I knew.

As a teenager helping my father install medical-imaging systems, I learned that invisible information does not become useful because someone believes it exists. Power, shielding, alignment, signal, calibration, and software have to work together. A new image is the product of an entire measurement architecture.

Recovering microbial chemistry at scale required the same systems thinking, but the object was alive and changing.

The winding road to water

My path did not look like a conventional route into drug discovery.

I studied environmental science at the Florida Institute of Technology and industrial engineering at Purdue. I entered high-technology manufacturing, built companies, created products and processes, and learned what it means to bring an invention through materials, equipment, quality, customers, and cost.

The successes were real. So were the reversals. A company I built reached the Inc. 500 five years in a row. Then the underlying technology changed and the market collapsed. We had to survive the sudden disappearance of a business we had become very good at serving.

That experience created a habit I did not fully appreciate until later: when an industry says a problem is impossible, ask whether the impossibility belongs to physics or to the installed process.

Our work in packaging and materials led toward biological feedstocks and algae. Algae promised renewable inputs. Harvesting them exposed the physical contradiction. The useful solids were microscopic and dilute; most of what the system moved was water. Conventional membranes fouled. Energy demand rose. Equipment that worked at laboratory scale could become uneconomic when multiplied across the volume required for a real process.

The solution had to respect mass balance.

How much water must move? How much biomass can be recovered? How gently? At what energy cost? What happens when the community contains organisms of different size, density, shape, surface charge, and fragility? How quickly does the chemistry change after separation?

The engineering eventually contributed to a $6 million ARPA-E program in 2009 around algal harvesting. But the most consequential output was not the original market application. It was the realization that a separation system capable of recovering microbial communities from very large volumes could become an instrument for natural-products discovery.

Scale was not merely something manufacturing solved after a molecule was found.

Scale could determine whether the molecule was findable.

The generalist's advantage - and burden

My route into the problem was broad enough to look unfocused from inside a specialized field.

Environmental science taught me to think in ecosystems and material flows. Industrial engineering taught me to think in bottlenecks, yield, energy, throughput, and failure modes. Manufacturing taught me that a promising material is not a product until equipment, quality, cost, and customers align. Entrepreneurship taught me that an elegant technology can die if it reaches the market at the wrong time or cannot survive the distance between proof and revenue.

Those experiences made me comfortable asking questions across boundaries. They did not make me a substitute for a natural-products chemist, microbiologist, pharmacologist, toxicologist, or computational scientist. The advantage of a generalist is not knowing more than specialists. It is being less committed to the borders among their fields.

The burden is translation.

The Whole Has No Department

Modern science advances by making questions small enough to answer. An engineer asks whether a living sample can be captured without destroying it. A microbial ecologist asks which organisms and interactions were present. A genomicist asks what capacities the community carried. An analytical chemist asks which molecular features can be measured. A natural-products chemist asks what structure can be isolated. A pharmacologist asks which targets or phenotypes move. A clinician asks whether any of it changes a patient's life. Each question is necessary. Each can be answered correctly. Yet no specialty is normally responsible for preserving the complete path from living context to molecule, from molecule to activity, and from activity to mechanism and outcome.

This is the specialist's paradox. **Specialization is how science earns depth, but isolation is how the whole can disappear.** Pure culture removes neighbors. Bulk extraction removes geography. Fractionation can remove synergy. A target assay removes the surrounding network. A clinical average can remove an individual chemical state. These methods are not mistakes; they are how variables become testable. The error comes when a controlled view is mistaken for the complete living system.

Ghost molecules are especially vulnerable at the handoffs. An ecologist can document a community event without preserving enough material for chemistry. A mass spectrometrist can label a reproducible feature "unknown" and leave it outside the biological story. A chemist can purify a molecule after the original activity vanishes. A pharmacologist can deprioritize a phenotype because no single target explains it. A genomicist can identify biosynthetic capacity without ever observing the resulting product. Every decision may be technically defensible. Together, they can turn one real biological event into a collection of orphaned clues.

The molecule has not disappeared from biology. **It has disappeared between disciplines.**

The answer is not a heroic generalist who pretends to know every field. It is a shared evidence architecture that allows specialists to remain specialists while keeping their answers attached to the same physical object and biological event: source, time, sample, genomic context, spectrum, structure, activity, negative result, target hypothesis, exposure, and phenotype. Specialists make each link trustworthy. Integration keeps the chain from breaking.

The ghost genome lives in the relationships. The observatory built to reveal it must therefore be interdisciplinary by design.

None, alone, describes the system.

The platform had to be built at their intersection.

This is also why the story is personal without being a memoir. My role is not to claim discovery of every scientific principle in these pages. It is to describe what became visible when engineering, microbial ecology, physical chemistry, and drug-discovery economics were forced into one problem.

One part per billion becomes matter

The arithmetic is simple enough to fit on a napkin.

At one part per billion, a liter of water contains roughly a microgram of a compound, depending on density and how the concentration is defined. A milliliter contains almost nothing. Across a million liters, the theoretical amount reaches the gram scale before losses.

Reality is harsher. The molecule may be divided among dissolved material, cells, particles, membranes, and complexes. Extraction is incomplete. Degradation begins. Interfering chemistry is abundant. The source is heterogeneous. A single number does not guarantee recoverable mass.

But the mass balance changes the experimental boundary.

At small scale, a feature may be detectable in a mass spectrometer but impossible to purify. Without enough material, the team cannot acquire the nuclear magnetic resonance data needed for structure, repeat biological assays, test stability, make a reference standard, or send material to another laboratory.

At larger scale, the same feature can become a fraction. A fraction can be retested. An active fraction can be separated again. Related spectra can be grouped into a molecular family. A tentative structure can be synthesized. The synthetic material can be compared with the natural sample and challenged across orthogonal assays.

Scale does not create biology that was absent. It can concentrate toxins, contaminants, and artifacts as readily as useful chemistry. It does not prove mechanism, safety, or therapeutic value.

It changes which questions can be answered.

Observability Is an Engineering Variable

A phenomenon does not become observable only because an instrument is more sensitive.

Observability depends on amount, preservation, separation, timing, and signal-to-noise. A molecule can be abundant across an ecosystem and absent from a vial. It can be detectable in a single injection and unavailable for the repeated work required to establish structure. It can survive collection and disappear during storage. It can be present in the sample and suppressed by more abundant chemistry during ionization.

Engineering changes those variables. Scale changes amount. Gentle collection changes preservation. Parallel extraction changes chemical coverage. Fractionation changes signal-to-noise. Time stamps and environmental metadata preserve context. Repeated aliquots make independent challenge possible.

The instrument is therefore larger than the mass spectrometer. It includes the pump, separator, container, cold chain, extraction, chromatography, metadata system, assay, and decision rule. A failure anywhere in that chain can make a real molecule scientifically invisible.

The objective is not to concentrate everything. It is to preserve enough reality that a trace can become a repeatable test.

What fails first in the field

A laboratory process lives inside controlled boundaries. A field process discovers how many of those boundaries were invisible.

Water carries suspended solids, dissolved salts, living cells, dead cells, extracellular polymers, debris, and organisms whose properties change during the day. Temperature moves. Wind redistributes the bloom. Rain changes dilution and runoff. A pump that handles clean water can damage fragile cells or clog. A membrane with excellent nominal pore size can foul when biological material coats its surface. A separator can perform differently as the community changes density or produces extracellular polymers.

The first failures are often not scientific. A fitting leaks. A hose collapses under suction. Power is interrupted. A preservation step takes longer than expected. A transport container is too warm. A process that looked continuous develops a bottleneck at one manual transfer.

Each operational failure can become a chemical bias.

If the system heats the sample, enzymes and oxidation change. If collection takes too long, the early material and late material represent different biological states. If the process selectively recovers larger cells, the library is not a community sample but a size-filtered one. If the preservation method favors one polarity or compartment, some chemistry disappears before extraction.

The engineering record therefore belongs in the scientific metadata.

This is one of the least glamorous and most important differences between discovery from a manufactured library and discovery from an ecosystem. A vendor compound arrives with a structure, amount, and purity. Environmental material arrives with a history that can determine what it contains.

A platform becomes credible when it can describe that history, reproduce key operations, quantify losses where possible, and admit where the sample is only an approximation of the original state.

A field instrument disguised as industrial equipment

The word *industrial* can mislead. It sounds like production after discovery - a large factory reproducing a known compound.

Industrial-Scale Microbiome Mining uses physical scale earlier. The ecosystem has already performed the cultivation. The goal is to recover enough of the community's output to make rare chemistry experimentally available.

This requires more than pumping volume.

A lake or bloom is heterogeneous across depth, light, temperature, nutrient gradients, wind, time, and species composition. A collection event is a sample of a moving ecology. The process must record location and time, preserve material quickly, avoid excessive mechanical damage, and create traceability from environment to fraction to spectrum to assay.

Collection itself changes the system. Cells experience pressure and shear. Oxygen changes. Light disappears. Enzymes keep working. Reactive intermediates can vanish or form. No method preserves the ecosystem perfectly.

The honest objective is not to claim an untouched snapshot. It is to understand the biases, standardize them, and preserve enough context that the physical material remains scientifically interpretable.

The workflow becomes a stack:

1. **Capture** the community with environmental metadata.
2. **Stabilize** material before uncontrolled chemistry erases the state.
3. **Extract** across complementary chemical conditions.
4. **Fractionate** the mixture into experimentally manageable libraries.
5. **Measure** using high-resolution mass spectrometry and related methods.
6. **Organize** features into families and link them to fractions, source, and activity.
7. **Confirm** structures with standards, synthesis, NMR, or orthogonal chemistry.
8. **Validate** function through exposure, target, phenotype, causality, and reproducibility.

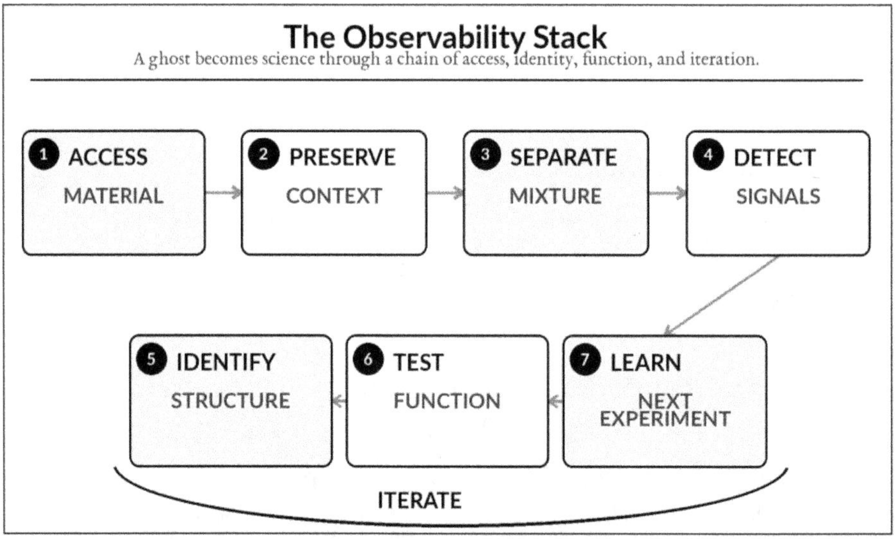

Figure 12. The observability stack: access, preserve, separate, detect, identify, test, and learn. Observability is a system property, not a single instrument.

No single step is the telescope. The system is the telescope.

When the machine becomes part of the hypothesis

In conventional laboratory thinking, the instrument measures the sample and should not change the scientific question.

At ecosystem scale, the collection machine inevitably becomes part of the hypothesis.

A filter selects by size and surface properties. A centrifuge selects by density. A membrane experiences pressure, fouling, and shear. Light changes while material moves through tubing. Temperature and oxygen change. Cells rupture. Dissolved molecules adsorb to polymers or metal. A preservation step freezes some reactions and permits others to continue until the temperature falls.

Each engineering choice edits the chemical record.

The correct response is not to claim that one process captures the ecosystem without distortion. It is to characterize the distortion and use multiple views.

A robust collection can separate cellular material from dissolved fractions, preserve time-matched controls, document water chemistry, and compare complementary capture methods. It can include field blanks, process blanks, replicate paths, and hold-time studies. It can ask whether a molecular family appears in the source water, increases in the captured material, or is created only after processing.

This level of discipline is familiar in analytical chemistry and environmental monitoring. What changes is the scale and purpose. The same process must produce enough material for discovery while preserving evidence about origin.

The engineering team therefore becomes part of the scientific team. A pump failure can reveal a sensitivity to shear. A clogged membrane can expose an unexpected biomass distribution. A change in freeze-drying behavior can indicate salts, polymers, or residual water that later affect extraction. Operational observations are not automatically mechanistic, but they can become hypotheses when linked to chemistry.

This was one of the cultural lessons of the work. Scientists often inherit a sample as though it simply arrived. Engineers often treat sample quality as a specification to be met. In microbiome mining, neither view is sufficient. The path from ecosystem to freezer is an experiment whose outputs include both material and uncertainty.

The machine does not merely collect the answer.

It determines which answer remains available to ask.

From biomass to wells on a plate

The distance between a recovered community and a screening library is a chain of decisions.

First comes stabilization. Freezing, drying, solvent exposure, pH control, and time can each preserve some chemistry and alter other chemistry. No universal method captures the entire metabolome. Complementary approaches are necessary because polar metabolites, lipids, peptides, and larger complexes behave differently.

Extraction separates material into chemical neighborhoods. A solvent that recovers hydrophobic compounds will miss many polar molecules. Strong conditions can degrade fragile structures. Gentle conditions can leave valuable chemistry behind.

Fractionation then reduces complexity. Chromatography divides an extract by retention behavior, charge, size, or other properties. The objective is not immediate purity. It is to create fractions simple enough for repeated assays and rich enough to preserve related chemistry.

The fractions are placed into wells and linked to analytical records. Each well becomes a physical address. A biological activity can be traced back to spectra and source. A molecular feature can be followed across adjacent fractions. When an activity disappears after separation, fractions can be recombined to test synergy. In some cases, available quantity will allow continuation of chasing a hidden active molecule.

This stage is where hidden chemistry becomes operationally compatible with modern drug discovery. A pharmaceutical assay does not screen a lake. It screens a plate.

The platform's job is to preserve as much ecological information as possible while translating the material into a form the laboratory can interrogate.

That translation creates losses. It also creates repeatability.

Without repeatability, a fascinating bloom remains a story. With repeatability, the material becomes a library that can be challenged by different assays over time.

The lake teaches humility

A microbial community is not a bag of species.

It is light gradients, nutrient exchange, predators, phages, minerals, temperature, and time. It is a history of who arrived, who died, what was consumed, and what remained. By the time material reaches a freezer, much of that history has already been compressed into chemistry.

Genomics can tell us which organisms and genes may have been present. It cannot reconstruct every chemical event that occurred before collection. The physical material contains an answer key written in a language we only partly read.

This is where the phrase *biological ground truth* became meaningful to me.

The ground truth was not a model prediction. It was the physical fact that a living community had made something. The feature had a mass, a fragmentation pattern, a place, a time, and a history before we assigned it a name.

The longer I worked with the material, the less plausible it seemed that the unknown chemistry of the human microbiome could be reduced to a tidy list of genes and predicted products. The human system is smaller in geography but greater in integration. Food arrives. Host enzymes cut. Microbes transform. Cells shed. Bile cycles. Drugs enter. Phages rupture cells. Oxygen changes across micrometers. The final molecular outputs are shaped by the whole moving system.

The lake did not prove the ghost genome.

It gave me permission to ask whether our smallest samples and cleanest cultures had trained us to underestimate it.

A library learns only when negative results remain visible

Discovery organizations naturally celebrate hits. The quiet majority of data are weak, ambiguous, or negative.

Those results are essential for artificial intelligence and for human judgment. A fraction that fails in one assay can still matter elsewhere, but the failure narrows the hypothesis. A predicted target that does not validate should reduce confidence in the model. A chemical family that repeatedly produces assay interference should be flagged rather than rediscovered as exciting noise.

Public scientific literature underrepresents negative experiments because journals, patents, and presentations reward novelty and success. Internal discovery systems can preserve them.

A physical library paired with a complete experimental history becomes more valuable as it accumulates failures. The data reveal which spectra correspond to unstable material, which fractions produce nonspecific cytotoxicity, which structural hypotheses recur incorrectly, and which assay classes are especially vulnerable to interference.

This is another reason the moat is not a molecule count. A large library with shallow annotations can be less informative than a smaller one connected to rigorous positive and negative evidence.

The goal is not to claim that every fraction is an asset.

It is to make every well capable of teaching the next experiment.

From one bloom to a reusable library

The public record of Biosortia's work includes research relationships with pharmaceutical, academic, and public institutions and the development of fractionated libraries from large microbial collections. Those collaborations mattered because a platform claim becomes useful only when material can move into the hands of independent scientists.

A library is more than a freezer full of extracts. It is an information architecture.

An Address Before an Explanation

A catalog does not have to wait for complete understanding to become valuable. We may not yet know exactly which organism made a molecule, whether another organism modified it, which meal supplied a precursor, why it appeared at that moment, or every tissue in which it acts. But once a reproducible molecule or molecular family has been physically captured, given an address, and linked to its spectrum, fraction, place, time, and observed activity, it becomes interrogable. Researchers can isolate or synthesize it, compare neighboring structures, expose cells and organisms to it, predict possible targets, test those predictions, and ask what changes when the molecule is added, removed, blocked, or altered. The unknown has crossed an important boundary: it is no longer merely an unexplained effect. It has become an experimental object.

That creates two maps at once. The first is a **molecule-to-activity map**: which structures produce which effects, at what dose, in what location, in what combinations, and for how long? The second is an **activity-to-target-and-mechanism map**: when a reproducible effect cannot be explained by known receptors, enzymes, pathways, or target combinations, the activity itself becomes a probe for biology our present maps do not yet contain. A known protein may have an unknown natural ligand. Several familiar targets may form an unfamiliar control pattern. A molecule may create an induced interface or act through a local cellular state that a conventional target screen would never reveal. Cataloging the chemical execution layer—including molecules more directly traceable to genomic pathways and context-created products of the ghost genome—therefore does more than add entries to a database. It expands the set of physical handles through which biology can be investigated and, eventually, responsibly influenced.

Each well or fraction should remain linked to source context, extraction conditions, chromatography, spectra, activity, and every later test. When a feature appears in several fractions, the relationships should be visible. When an assay produces a weak or negative result, the result should remain part of the record. When a structure is proposed, confidence should be explicit. When the material is exhausted, resupply and synthesis status should be known.

This is difficult and unglamorous work. It is also where a discovery platform becomes cumulative rather than anecdotal.

In the 2026 Muskegon collaboration described in the prologue, molecular networking and product-ion queries helped reveal cyanopeptide families across a prefractionated bloom library. The scientific importance was not the raw number of peaks. It was the ability to move from physical collection to fractions, from fractions to spectra, from spectra to families, and from families to isolated and characterized molecules.

The path is repeatable in principle.

Repeatability turns a strange sample into an instrument.

External scientists change the platform

A proprietary discovery system can become intellectually closed if every result is interpreted by the team that built it. External collaboration provides a different form of calibration.

Publicly described Biosortia relationships have included pharmaceutical and academic researchers working in infectious disease, cancer, natural-products chemistry, and environmental microbiology. The value of those relationships was not endorsement of every platform claim. It was exposure to assays, questions, and standards the originating team did not control.

A cancer biologist sees a fraction differently from an ecologist. An analytical chemist challenges identification confidence. A pharmaceutical scientist asks whether the material can be resupplied and whether the activity survives purification. A clinician asks whether the proposed effect can ever be measured in a patient.

The collaborations can be frustrating because they reveal how far a discovery remains from translation. They also prevent the platform from becoming a self-confirming story.

The strongest external result is not praise. It is a reproducible finding that survives another laboratory's methods and skepticism.

The first molecular city map

The first time a large untargeted dataset resolves into a network rather than a list, the effect is disorienting.

A conventional table places one feature beneath another: mass, retention time, intensity, possible formula, possible annotation. The rows can number in the tens of thousands. Most names are blank. The unknowns feel like clerical failure.

A molecular network changes the visual grammar. Spectra with related fragmentation patterns cluster. Known compounds become small islands of annotation surrounded by families of unnamed neighbors. Repeated chemical motifs appear across fractions. A feature seen in one sample can connect to analogs found in another season, location, or biological activity.

The map begins to look less like debris and more like a city at night.

A few streets are labeled. Most buildings are dark. Yet the pattern of roads reveals neighborhoods before we know who lives in them.

For an industrial-scale library, that organization is essential. Physical fractions make it possible to return to a neighborhood. Environmental metadata tells us where it appeared. Bioassays tell us whether one branch of the network follows a phenotype. High-resolution spectra and computation suggest which modifications distinguish family members. Synthesis and standards eventually turn a street address into an identity.

The deepest value is not the novelty count. A claim such as *tens of thousands of unknown features* can impress a presentation and say almost nothing about development value. The network becomes useful when it guides decisions:

- Which family is reproducible across independent collections?
- Which members track with activity?
- Which fraction contains enough material for isolation?
- Which analogs appear to be oxidation products or processing artifacts?
- Which structural hypothesis would most efficiently distinguish the candidates?
- Which negative assay removes an entire branch from consideration?

This is where computation stops being decoration around a natural-product library. It becomes the navigation system for physical material.

The city map also changes the emotional relationship to the unknown. Blank names no longer mean absence of information. They mean that relational information arrived before full identity.

That is often how a frontier first becomes legible.

The First Lighthouse Problem

Every new discovery instrument eventually faces a fair and uncomfortable question: What result can the existing system evaluate?

Pharmaceutical organizations advance assets, not ontologies. Investors and partners can value a confirmed structure, reproducible assay, manufacturing path, safety plan, and clinical opportunity. They cannot easily price the option value of thousands of unresolved molecular families.

The instrument and the application occupy different stages, but the organization building the instrument must survive both. That creates two clocks. The scientific clock measures how long evidence genuinely requires. The survival clock measures how long the team can continue before capital, attention, or credibility runs out.

The answer is a lighthouse asset: one program valuable on its own terms and clear enough to illuminate the discovery architecture behind it. It does not need to prove every ghost-genome claim. It needs to show that chemistry missed by conventional starting points can become a credible development candidate.

The strongest first lighthouse may not be the largest market or the most dramatic disease. It may be a local therapy with a measurable phenotype, a tractable supply route, and an evidence package that can mature quickly. Gut, skin, wound, anti-infective, or other geofenced settings can make mechanism and safety easier to separate.

The gates should be explicit:

1. the physical feature and molecular family are reproducible;
2. identity and stereochemistry are confirmed at the level required for the claim;
3. activity survives purification, dose-response, counterscreens, and orthogonal assays;
4. resupply and manufacturing are credible;
5. exposure is achievable in the intended biological geography;

6. early safety and off-target liabilities are manageable;
7. the differentiated phenotype or mechanism survives comparison with a meaningful standard.

A platform must preserve unusual possibilities while killing weak leads aggressively. Unusual is not the same as valuable. The grand narrative earns permission to look; the asset earns the next dollar.

Biosortia appears in this book as one firsthand attempt to solve that problem, not as the subject of the book and not as proof that the problem has been solved. The broader case does not depend on one company surviving. The access and observability gap remains whoever builds the instrument.

Catching molecules before they vanish

Physical scale solves only one form of invisibility: insufficient material.

Other ghost molecules are lost because they are spatially confined, rapidly turned over, strongly bound, unstable, poorly ionized, or created only during a brief state. Reaching them requires additional instruments.

Targeted metabolomics measures known compounds precisely. Untargeted metabolomics surveys a broader field but produces many unresolved features. Molecular networking groups related spectra even when identities are missing. Spatial metabolomics preserves geography. Stable-isotope tracing follows atoms through host, microbial, and dietary pathways. Activity-guided fractionation follows phenotype through separation. Chemoproteomics can reveal engaged protein families. Time-resolved sampling distinguishes cause, compensation, and debris.

The stack is expensive. It is also cheaper than advancing persuasive false positives through years of development.

A reproducible feature is the beginning, not the end. Full identification may require exact mass, fragmentation, isotope pattern, retention behavior, derivatization, NMR, stereochemical analysis, synthesis, and comparison with an authentic standard. Biological validation requires dose-response, controls, orthogonal assays, exposure measurements, and intervention in a relevant system.

The new telescope must be paired with new humility.

The physical moat

Artificial-intelligence companies often describe algorithms, compute, and proprietary models as their competitive moat. Those assets matter. In biological discovery, the deeper moat may begin before the data become digital.

Competitors can buy processors. They can reproduce a published architecture. They cannot instantly recreate a physical collection, its ecological history, the exact fractions, the negative assays, the operator knowledge, the spectral families, and the chain of decisions that linked one unusual feature to a useful experiment.

The data advantage is material before it is computational.

That does not make the material sacred. A library can become a warehouse of untested fractions. The value emerges only through a closed loop: physical evidence guides computation; computation prioritizes the next experiment; the experiment updates the interpretation; and the updated interpretation changes what is collected, separated, synthesized, or tested next.

The lake is therefore not the opposite of artificial intelligence.

It is the part of the intelligence system that touches reality.

The next story introduces another instrument built around the same reversal. Instead of forcing an uncultured microbe to accept the laboratory, a small device returned the microbe to its environment and allowed nature to complete the experiment.

CHAPTER 8

The Chip That Returned the Microbe Home

When the Environment Became Part of the Experiment

STORY 8

THE CHIP THAT RETURNED THE MICROBE HOME

When the Environment Became Part of the Experiment

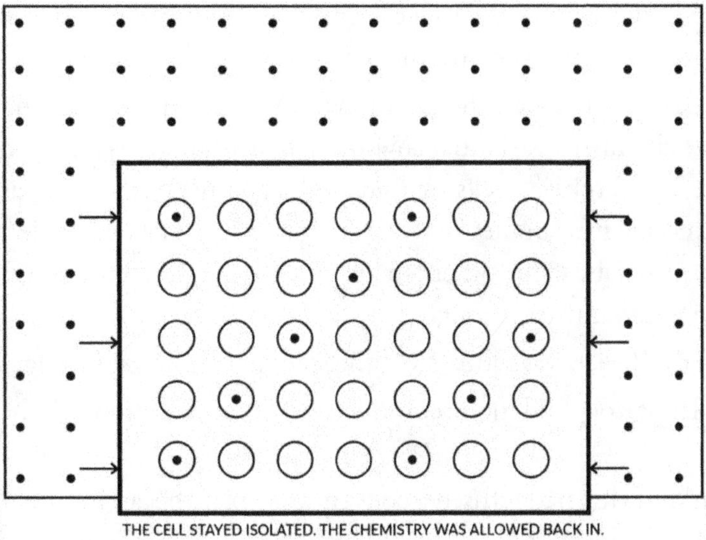

THE CELL STAYED ISOLATED. THE CHEMISTRY WAS ALLOWED BACK IN.

THE TURNING POINT

Isolation was preserved. Native chemical context was restored.

The Great Plate Count Anomaly began as an embarrassment visible through a microscope.

Count the cells in an environmental sample directly and the number is enormous. Spread the same sample on standard laboratory media and wait for colonies; only a small fraction appears. The cells were present. The plate recovered the minority willing to accept its food, oxygen, temperature, isolation, and timing.

Microbiology knew the discrepancy for decades. The standard response was to improve media, vary culture conditions, or use molecular methods that avoided cultivation. Each approach expanded the visible world. The uncultured majority remained.

Slava Epstein, Kim Lewis, and colleagues changed the question.

Instead of asking why an environmental organism failed to grow in the laboratory, they asked how to let it grow while remaining exposed to the chemistry of its natural environment.

The device was called the isolation chip, or iChip. It contained many miniature chambers. A diluted environmental sample was arranged so that individual microbial cells could occupy separate compartments. Semipermeable membranes kept the cells physically isolated while allowing nutrients, signals, and other small molecules from the surrounding environment to diffuse through.

Then the device was returned to the soil or other native setting.

The laboratory did not reproduce nature's conditions. It borrowed them.

Organisms that had refused a conventional plate began to form colonies within the chambers. Some could later be domesticated and grown under laboratory conditions after the initial environmental incubation. The iChip did not solve unculturability. It changed the success rate enough to turn an old anomaly into a discovery method.

One of the organisms recovered through this approach, *Eleftheria terrae*, produced a compound named teixobactin. The molecule showed activity against several Gram-positive pathogens and drew intense attention because investigators did not detect resistance under the experimental conditions reported in the initial study.

Teixobactin was the fish. The iChip was the fishing rod.

The deeper lesson was not that every uncultured organism contains a revolutionary antibiotic. It was that the environment was part of the organism's operating system.

Hundreds of tiny rooms

The iChip looks modest compared with the question it addresses.

The device contains many small chambers or wells. Environmental material is diluted so that individual cells, or very small numbers of cells, occupy separate spaces. Semipermeable membranes hold the cells in place while allowing molecules from the surrounding environment to diffuse through. The assembled device is returned to the source environment.

The organism is physically isolated without being chemically exiled.

This is an elegant inversion of ordinary culture. A standard plate asks the cell to accept a laboratory's chosen nutrients, oxygen, neighbors, surfaces, and signaling conditions. The iChip allows the laboratory to control separation while borrowing much of the environment's chemical context.

The method does not reproduce every ecological interaction. A cell may still lose direct contact with a partner. The membrane has permeability limits. The chamber changes spatial structure. Some organisms remain dependent on conditions the device does not preserve. Yet the increase in cultivation success showed that many cells labeled unculturable were not incapable of growth. They were incompatible with the experiment we had designed.

That distinction is intellectually important.

Unculturable sounds like a property of the organism.

Not cultured under these conditions describes the state of a method.

The same correction applies across discovery. A molecule may be called undetectable when the extraction does not recover its class. A target may be called undruggable when the library lacks the appropriate shape. A community effect may be called irreproducible when the assay removes the substrate or neighbor that creates it. A disease may be called idiopathic when the causal layer has not been measured.

Good instruments move failure from the object back into the method where it can be redesigned.

From a cell in soil to a colony in the laboratory

The iChip did not eliminate isolation. It changed the order.

In a conventional workflow, a cell is removed from its environment and asked to grow immediately on defined laboratory media. If it fails, the organism disappears from the research pipeline.

In the iChip workflow, cells are separated into small chambers while the device remains in contact with native environmental chemistry. Nutrients and signaling molecules diffuse through the membranes. The organism can divide in a setting closer to the one in which it evolved.

After colonies form, researchers attempt *domestication* - gradually transferring the organism to laboratory conditions. Some microbes that could not initiate growth on a standard plate become cultivable after the environmental stage, perhaps because the initial population size, physiological state, or access to growth factors changes.

This is a profound change in failure logic. The organism does not have to prove laboratory independence at the first step. It is allowed to reveal itself before being asked to survive the reduction.

The method also creates traceability. A colony can be linked to a specific chamber, isolated, sequenced, fermented, and rescreened. That direct path from organism to product is one of cultivation's great advantages.

The limitation remains equally important. The chemistry produced after domestication may differ from the chemistry produced in situ. The organism can stop making the compound when the environmental trigger is removed. A neighboring species may have supplied a precursor. The product that led to discovery may be only one member of a family expressed under changing conditions.

Cultivation retrieves the producer. It does not automatically retrieve the original state.

Context is not contamination

Laboratory science depends on control. Remove variables. Isolate the organism. Define the medium. Standardize temperature, oxygen, and time. The reduction allows causal inference.

But context can be removed so thoroughly that the function disappears.

A microbe may depend on nutrients supplied by a neighbor, signals that indicate population density, detoxification performed by another species, a mineral surface, a host molecule, a phage pressure, or a concentration gradient. In pure culture, the organism is not simply the same organism under cleaner conditions. It can be a different physiological state.

The iChip embodies a disciplined compromise. It isolates the cell enough to observe growth while preserving environmental exchange. The membrane does not recreate the community. It allows the missing chemistry to cross.

That reversal is central to the ghost genome.

The conventional experiment treats environment as noise around the biological object. The new experiment asks which part of the environment is required for the object to become what it is.

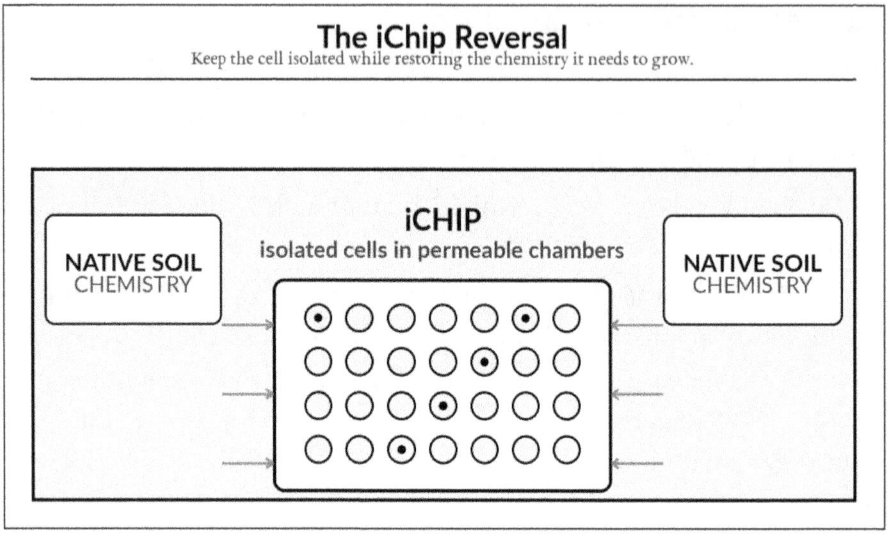

Figure 13. The iChip reversal: isolate the cell physically while allowing native chemical context to diffuse through.

What teixobactin actually demonstrated

Teixobactin became a global headline because it attacked important Gram-positive bacteria and investigators did not detect resistance under the conditions of the initial experiments.

The molecule binds highly conserved cell-wall precursors, including lipid II and related molecules required for building peptidoglycan and teichoic acids. These targets are not proteins encoded by one easily mutated gene. They are essential chemical building blocks. That mechanism offered a plausible reason resistance might be more difficult to evolve.

More difficult is not impossible.

Resistance can arise through altered uptake, efflux, cell-envelope remodeling, sequestration, community protection, or mechanisms that do not change the direct binding target. Clinical use introduces population sizes, exposures, tissues, and evolutionary pressures beyond an initial laboratory study.

The careful statement is that the discovery revealed a valuable molecule and a mechanism with an encouraging resistance profile. It did not repeal evolution.

The greater platform proof was independent of whether teixobactin ultimately becomes a widely used medicine. A method built around environmental access recovered an organism and chemistry that the conventional plate had missed.

The instrument changed the candidate distribution.

What teixobactin did not promise

The 2015 teixobactin paper attracted extraordinary attention in part because the authors reported no detectable resistance under the experimental conditions used with certain pathogens.

The phrase was easy to enlarge into *an antibiotic to which resistance cannot evolve*. That was not the demonstrated claim.

Teixobactin binds conserved cell-wall precursors rather than a single mutable protein pocket, which can raise the barrier to resistance. The relevant target architecture is one reason the molecule was scientifically important. But bacteria can evade drugs through more than point mutation: altered permeability, efflux, community protection, enzymatic inactivation, changes in physiology, or acquisition of new functions can all matter. Clinical resistance is an ecological and evolutionary outcome, not a property that can be declared impossible from one set of experiments.

This caution does not diminish the iChip result. It improves it.

The true proof of concept was access. A bacterium that standard cultivation had missed produced a chemically distinctive antibiotic with a compelling mechanism. The method expanded the set of organisms available for study and thereby expanded the chemistry available for discovery.

For the ghost-genome argument, that is enough.

A new access method does not need every molecule it reveals to be superior. It needs to produce starting points the old method systematically failed to produce, then allow ordinary pharmacology to determine which ones are useful.

This is the same standard that should be applied to industrial-scale microbiome mining. The platform should not be judged by the number of unknowns or by a promise that nature solved resistance. It should be judged by prospective yield: does the method deliver reproducible, structurally confirmed, biologically relevant starting points at a rate and quality that justify the access cost?

A telescope succeeds by revealing a world. It does not guarantee that every object in the field of view is habitable.

One organism versus a community's chemistry

The iChip and industrial-scale microbiome mining approach the access problem from different directions.

The iChip begins with cells. It gives individual organisms a way to grow under environmental influence, then attempts to bring the productive strain into a tractable culture. This can reveal the producer, enable resupply, and connect a molecule to a genome.

Industrial-scale community recovery begins with chemistry already produced in situ. It does not require the individual producer to be known at the moment of capture. It can preserve products arising from competition, cooperation, phage events, or sequential transformations that isolated culture may not reproduce.

Each method gains something and loses something.

Cultivation provides an organism, a renewable source, and a clearer genetic path. It can lose the native community state.

Community harvesting preserves an ecological output and can make low-abundance material physically available. It can lose direct authorship and create a difficult resupply problem.

The two are complementary. An active molecular family discovered in community material can guide cultivation, genome mining, co-culture, heterologous expression, or synthesis. An unusual organism recovered by iChip can be returned to defined communities to ask how its chemistry changes in context.

The new telescope is a stack of instruments because no one instrument preserves every truth.

A biosynthetic gene cluster is a promise

Genomics transformed natural-products discovery by revealing biosynthetic gene clusters - neighboring genes encoding enzymes, regulators, transporters, and resistance functions associated with specialized metabolites.

Tools such as antiSMASH can identify clusters that appear to build nonribosomal peptides, polyketides, ribosomally synthesized and modified peptides, terpenes, and other families. Databases such as MIBiG connect experimentally validated clusters to products. Researchers can prioritize unusual pathways, activate silent genes, and express clusters in new hosts.

A biosynthetic gene cluster is one of the most powerful priors in modern discovery.

It is not a photograph of the final chemical world.

One cluster can produce a family rather than one compound. Enzyme modules may accept alternative substrates. Tailoring reactions may be incomplete. Regulatory conditions change flux. Culture conditions yield one analog while the native community yields another. Transport and degradation determine what accumulates.

Many clusters are silent or weakly expressed under standard conditions because the relevant trigger is absent. The organism has no reason to spend energy making an expensive molecule when there is no competitor, host signal, nutrient limitation, oxidative stress, quorum cue, or phage pressure.

The genome contains capacity. The environment decides whether capacity becomes chemistry.

Silent clusters and the temptation of prediction

Once genomes revealed large numbers of unassigned biosynthetic clusters, a seductive conclusion followed: perhaps the hidden chemical world could be predicted directly from sequence.

In some cases, prediction is impressively informative. Modular nonribosomal peptide synthetases and polyketide synthases contain domains whose order and specificity suggest portions of the backbone. Tailoring enzymes indicate methylation, glycosylation, oxidation, halogenation, or ring formation. Resistance genes can hint at biological target.

The challenge is that biosynthesis behaves less like deterministic compilation than the clean diagram suggests.

Enzyme specificity can be broader than expected. Modules can be skipped. Starter units vary. Tailoring can be incomplete. Pathways branch. Products are exported, degraded, or modified by other organisms. Genome assemblies can split or misassemble large clusters. Annotation tools are trained on the characterized examples, which are a biased subset of nature's chemistry.

A predicted structure is therefore a hypothesis with a confidence level, not a molecule.

The danger grows when computational systems generate a plausible product and an in silico spectrum that appears to match an unknown feature. Shared assumptions can make the pair look more certain than the evidence warrants.

The physical standard remains decisive. Does the predicted compound reproduce retention time, fragmentation, ion mobility, NMR, stereochemistry, and biological activity?

Sequence narrows the suspects. Chemistry identifies the material.

The final molecule may have several genomes

Consider the bile-acid circuit. Human genes produce primary bile acids. Different microbial genomes encode deconjugation and transformation steps. The final pool depends on community composition, diet, circulation, and tissue state. There is no single microbial gene cluster for the complete output.

Consider a dietary polyphenol. A plant genome made the precursor. Food processing changed it. One microbe removed a sugar. Another opened a ring. The host conjugated the product. The kidney determined exposure.

The molecule is real. Its authorship is distributed.

The cryptic-assembly hypothesis generalizes the pattern. It asks whether some uncharacterized signaling molecules arise through transformations crossing species and kingdoms, using substrates supplied by food, cell turnover, and neighboring metabolism.

If so, genome mining will provide essential pieces without ever containing the whole route.

The limitation is not a weakness of genomics. It is a boundary condition. Genomics measures capacity. Transcriptomics moves closer to current intent. Proteomics measures machinery. Metabolomics measures products and intermediates. Spatial methods preserve geography. Isotope tracing assigns atoms. Perturbation establishes necessity.

The layers can disagree.

A gene can be present and silent. RNA can be present while translation is limited. An enzyme can be present without substrate. A metabolite can persist after expression stops. A neighboring organism can consume the product immediately. A host can clear it before it accumulates.

This is not noise around the true genomic signal.

It is the system.

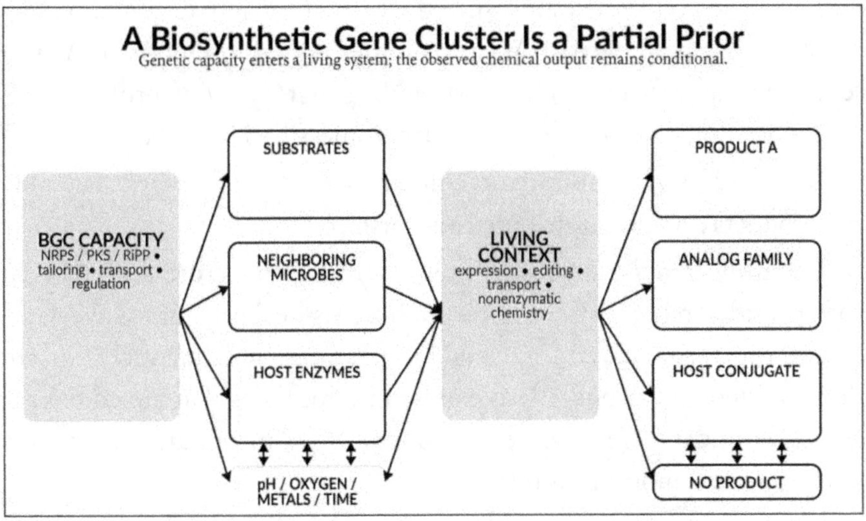

Figure 14. A biosynthetic gene cluster is a partial prior: genetic capacity enters living context and can yield several products - or none.

A distributed pathway in practice

Bile-acid metabolism shows what a distributed pathway looks like without requiring any speculative chemistry.

The host liver synthesizes primary bile acids from cholesterol and often conjugates them before secretion. The molecules enter the intestine as digestive agents and signals. Microbial enzymes can remove conjugates, alter hydroxyl groups, change stereochemistry, and perform other transformations. The edited molecules are reabsorbed, sensed by host receptors, modified again, and returned through enterohepatic circulation.

No single bacterial biosynthetic gene cluster owns the final pool.

The host supplies the starting scaffold and transport system. Different microbes contribute different enzymatic steps. Diet changes the demand for bile and the ecology of the intestine. Transit time changes exposure. The liver and kidney influence clearance. The final signaling mixture is a system output.

A similar relay occurs with many dietary polyphenols. The plant made the precursor. Food processing altered its structure and accessibility. One microbial enzyme removes a sugar. Another organism cleaves or reduces the remaining scaffold. The host adds sulfate or glucuronide. The circulating conjugate becomes the molecule measured in blood.

These pathways establish the conservative core of cryptic assembly: final products can have distributed authorship.

The stronger hypothesis asks how far this architecture extends. Can short peptides released from food or host tissue be trimmed, acylated, oxidized, ligated, cyclized, or metal-bound through sequential host and microbial steps? Can phage lysis expose intermediates that are edited at a mucosal boundary? Can transient complexes carry information even when no new covalent molecule is formed?

The experimental strategy is the same whether the answer is common or rare.

Label the precursor. Remove one participant. Observe the intermediate. Restore the participant. Confirm the product. Measure the exposure. Test the phenotype. Repeat in a second system.

Distributed authorship is not an excuse for vague mechanism.

It is a map for assigning mechanism one step at a time. It is a prior for AI training data for the universal control layer and humanities' future.

The chemistry no cluster owns

Some products will remain outside even perfect gene-cluster prediction because no single cluster owns them.

Host enzymes commonly conjugate microbial metabolites with sulfate, glucuronide, glycine, or other groups. Microbes can remove those conjugates and perform another transformation. A dietary precursor can be required. A nonenzymatic oxidation can complete the active form. A metal can organize the molecule into a functional state.

The pathway is distributed across genomes and chemistry.

Mapping such a pathway requires perturbation rather than annotation alone.

Start with a reproducible feature. Use isotope-labeled substrate to trace atoms. Remove one microbial species and determine whether the feature disappears or an intermediate accumulates. Knock out a candidate enzyme. Add the species or enzyme back and test rescue. Block host conjugation. Change pH, oxygen, or metal availability. Synthesize the proposed product and compare analytical and biological behavior.

Each experiment assigns part of the authorship.

The completed map may look like a pathway diagram, but the arrows cross organisms, tissues, and compartments. The environment is no longer a box drawn around the pathway. It is one of the reactants.

Nonenzymatic chemistry deserves special caution. Oxidation, hydrolysis, glycation, nitrosation, and metal-catalyzed reactions occur in living systems and during sample processing. A product detected after extraction may be an artifact rather than an in vivo signal.

Controls must ask when and where the molecule formed. Rapid quenching, isotope experiments, alternative extraction conditions, and in situ methods can distinguish biology from handling.

A ghost molecule becomes credible when its route can be reconstructed, not merely imagined.

Awakening silent chemistry

Researchers have developed many strategies to activate silent biosynthetic capacity. Change nutrients. Add competitors. Co-culture organisms. Alter epigenetic regulation. Introduce stress. Express a cluster in a heterologous host. Use CRISPR-based activation. Search for pathway-specific regulators.

Each technique is a way of asking the genome a different question.

The iChip's question is environmental: What will this organism do if the molecules it normally receives are allowed to reach it?

Community harvesting asks a later question: What did the ecosystem actually make before we separated it?

Spatial metabolomics asks: Where did the product accumulate?

Isotope tracing asks: Which atoms came from which source?

Artificial intelligence can help connect the answers, but only if the experiments preserve enough provenance to distinguish prediction from observation.

The danger of the clean story

Discovery narratives are often rewritten backward. A soil sample produced an organism. The organism produced a compound. The compound hit a target. The target explained a disease. The medicine followed.

The actual path is usually messier. Multiple extracts are screened. Activities disappear. Structures are misassigned. Supply fails. Mechanisms change. Programs are paused. A molecule finds a different indication from the one that justified its first assay.

The clean story is useful for teaching. It can also hide which part of the system created the discovery.

Teixobactin's significance was not only its activity. It was the proof that changing access could reveal chemistry absent from the conventional pipeline. The iChip made a previously uncultured organism available because the device treated native chemistry as an experimental requirement rather than an uncontrolled nuisance.

The same principle applies inside the human body. A microbial molecule may require mucus, diet, a neighboring species, host enzyme, or local oxygen state. Removing context can make the pathway look genetically silent or chemically nonexistent.

The correct response is not to abandon reductionism. It is to reduce the system in stages while checking that the function survives each reduction.

When the activity disappears, the disappearance is data.

A practical context ladder

Context can become an excuse for experiments that are too complex to interpret. A disciplined program should add complexity in steps.

Begin with the simplest system in which the activity appears. Test the isolated molecule where possible. Add the proposed target, then cells, then defined co-cultures, then tissue architecture, then a more complete community. At each stage, ask whether the effect survives and which new variables become necessary.

The reverse ladder is equally useful. Begin with the intact active system and remove components one at a time. Fractionate the chemistry. Remove a species. Block an enzyme. Change a substrate. Separate contact-dependent from diffusible effects.

When the phenotype disappears, the missing component becomes a clue.

This approach protects against two opposite errors. Reductionism can remove the mechanism before it is identified. Systems biology can preserve so many variables that causality becomes impossible.

The goal is not maximum complexity.

It is the minimum sufficient context.

The iChip found that native chemical exchange was part of the minimum sufficient context for many organisms to begin growing. The ghost-genome program must discover the analogous minimum context for each molecule or phenotype.

Preserving context without preserving confusion

There is a danger in every call to study biology in context: context can become a reason not to simplify anything.

A whole ecosystem contains thousands of variables. If every result is explained by unspecified complexity, no causal knowledge emerges. The objective is not to choose between reductionism and context. It is to use them in sequence.

A productive workflow can begin with an intact system to discover a reproducible effect, then remove dimensions until the effect disappears.

A community fraction changes a host-cell phenotype. The fraction is separated. Activity follows two molecular families. Removing one family weakens the effect; recombining both restores it. A defined consortium reproduces production. One organism supplies a precursor, another performs a modification. A host enzyme completes the product. The synthetic standard reproduces the spectrum and part of the phenotype. The remaining difference reveals either synergy or an incorrect structural assumption.

Each reduction is an experiment, not a declaration that the isolated part is the whole explanation.

The iChip teaches the reverse sequence as well. When isolation destroys growth, add back environmental features in controlled steps: spent medium, siderophores, signaling molecules, neighboring strains, surfaces, oxygen gradients. The missing dependency becomes visible through rescue.

These two directions - simplify from context and rebuild toward context - should meet.

That meeting point is a causal model rich enough to reproduce the phenomenon and simple enough to test.

The ghost genome will not be decoded by preserving every variable forever. It will be decoded by learning which variables cannot be removed without losing the chemistry.

The environment becomes part of the assay

A mature ghost-genome program would use context systematically rather than romantically.

It would define environmental variables, record them, and perturb them one at a time where possible. It would compare pure culture, co-culture, native community, and reconstructed consortia. It would test whether a feature can be reproduced with a defined substrate and enzyme set. It would distinguish a molecule created in the source from one created during extraction. It would build reference standards and negative controls.

The objective is not to preserve nature untouched.

The objective is to learn which parts of nature are causal.

The iChip returned a cell to its environment long enough for the organism to reveal itself. The lake-scale approach recovered a community's chemistry before demanding that every producer reveal itself. Both methods begin with the same act of humility:

Maybe the failure is not in the organism.

Maybe the experiment removed what the organism needed to be visible.

The Genome Predicts Capacity; the Assay Observes Execution

The iChip story provides the cleanest possible distinction between what a genome allows and what a living system actually does.

The uncultured organism already possessed the genes required for growth and chemistry. Standard laboratory conditions did not activate the necessary state. Returning the cell to environmental diffusion supplied missing nutrients, signals, partners, or physical conditions. The same genome produced a different observable reality.

Biosynthetic-gene-cluster prediction is therefore a searchlight, not a final inventory. It can identify promising capacity, suggest a scaffold, and guide which spectra to seek. The physical assay tells us whether the capacity executed under the condition that matters.

For the ghost genome, the complete loop is circular. Sequence guides chemical observation. Chemistry reveals which sequence or community to perturb. Perturbation changes the state. The changed state teaches the next model.

A good discovery system refuses to choose between genomics and chemistry. It asks each layer the question it is capable of answering.

The next story turns to artificial intelligence, the most powerful new instrument in biological science. Its greatest success came not from ignoring the accumulated structure of nature, but from learning a problem whose representation, data, and physical answer were unusually well aligned.

PART IV

Intelligence and Translation

Two stories about turning hidden biology into something medicine can use.

• • •

CHAPTER 9

The Day AI Solved the Right Problem

What AlphaFold Teaches - and What It Does Not

STORY 9

THE DAY AI SOLVED THE RIGHT PROBLEM

What AlphaFold Teaches - and What It Does Not

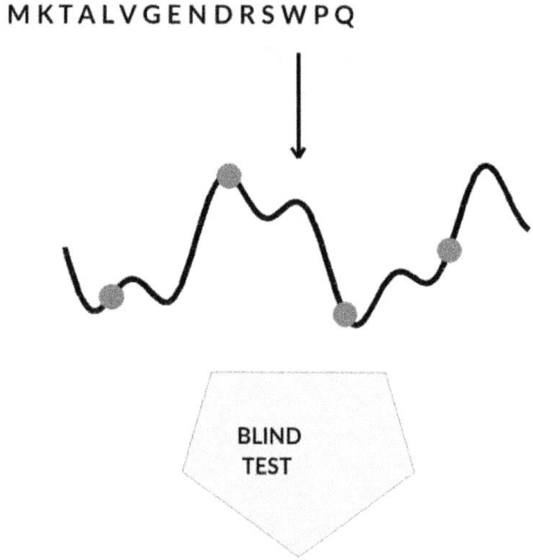

THE TURNING POINT

AI became credible when prediction met a protected physical answer.

For decades, protein folding was described as one of biology's grand challenges.

A protein begins as a linear chain of amino acids. Its function depends on the three-dimensional shapes and assemblies the chain can adopt. In principle, the sequence contains the constraints. In practice, the number of possible conformations is enormous, interactions occur across many scales, and the cellular environment matters.

Experimental structural biology transformed science, but it was difficult and slow. X-ray crystallography, nuclear magnetic resonance, and cryo-electron microscopy could reveal atomic or near-atomic structures, yet each required specialized work and not every protein cooperated.

The Critical Assessment of protein Structure Prediction, or CASP, created a disciplined test. Researchers received protein sequences whose experimental structures had not yet been publicly released. Teams submitted predictions. The predictions were compared with ground truth after the structures became available.

CASP did something innovation systems often fail to do: it defined the problem, protected the answer, and forced competing methods to confront the same evidence.

Progress came gradually. Then, in the 2020 CASP14 competition, AlphaFold 2 produced predictions that approached experimental accuracy for many targets. The achievement was not a press-release illusion. It appeared in a blinded, community-run benchmark against physical structures.

AlphaFold did not solve every question about protein dynamics, disorder, complexes, mutations, membrane environments, or function. It changed what scientists could reasonably expect from sequence-to-structure prediction.

In 2024, AlphaFold 3 extended the ambition. A new diffusion-based architecture predicted joint structures of complexes containing proteins, nucleic acids, small molecules, ions, and modified residues. The model reported improved performance across several classes of biomolecular interactions compared with earlier specialized tools.

The breakthrough is one of the strongest arguments against casual pessimism about artificial intelligence in biology.

It is also routinely used to support claims AlphaFold itself does not make.

A benchmark that protected the field from its own excitement

Protein-structure prediction had endured waves of confidence before AlphaFold. Energy functions, fragment assembly, comparative modeling, and machine learning all produced advances. The field also learned how easy it was to overestimate progress when methods were tested on structures already known, closely related to training examples, or selected after the fact.

CASP created a recurring public reckoning. Experimental groups supplied sequences for structures that had been solved but not released. Predictors could not tune their answers after seeing the target. Independent assessors compared models with the hidden structures and scored different aspects of accuracy.

The benchmark was not perfect. CASP targets did not represent every class of protein or every biological question. But it gave the community a shared clock and a protected answer key.

AlphaFold 2's performance at CASP14 mattered because the system crossed a threshold under those conditions. Predictions for many targets approached the accuracy expected from experimental methods, especially in the protein backbone. Confidence estimates helped users distinguish regions the model trusted from those it did not.

The achievement emerged from several ideas working together. Evolutionary relatedness among sequences provided constraints: when two residues change together across species, they may be physically or functionally linked. Attention-based neural networks allowed information to move among sequence positions and spatial relationships. Iterative refinement let the model update the structure as its internal representation improved.

The details were technically new. The strategic pattern was older: use a large physical archive, encode the right constraints, and submit the result to an external measurement.

This is the opposite of synthetic confidence.

The model's reputation did not come from one company scoring its own predictions with another internal model. It came from surviving a community test whose answers had been withheld.

Drug discovery needs more benchmarks with that character. Prospective target predictions should be tested before the biology is known. Generated molecules should be compared under matched budgets and assays. Failed compounds should remain visible. Whole-program time, cost, and attrition should matter, not only retrospective enrichment on curated datasets.

Without protected answers, the field can select the success story after seeing the outcome.

Why AlphaFold was the right kind of problem

AlphaFold's success did not come from ignoring biology. It came from representing biology in a form a learning system could use.

Protein structure prediction had several advantages.

The input was definable: amino-acid sequence, often enriched by evolutionary relationships among related sequences.

The target was physical: an experimentally determined three-dimensional structure.

The training archive, the Protein Data Bank, was imperfect and biased but contained a large collection of structures accumulated over decades.

The geometry was constrained by chemistry. Bond lengths, angles, residue interactions, steric limitations, and evolutionary conservation restricted the answer space.

The benchmark was external. CASP made it difficult to confuse confidence with correctness.

The model still had to discover an extraordinary mapping. But the problem had a shape.

Drug discovery is more open-ended.

A molecule can bind a protein beautifully and still fail because it does not reach the tissue, is metabolized too quickly, blocks an ion channel, injures the liver, triggers an immune response, forms a reactive metabolite, cannot be manufactured, or produces no meaningful change in human disease. A disease can contain multiple biological states under one diagnostic label. The target can be correct and the therapeutic hypothesis wrong.

Structure is a powerful layer.

It is not the whole clinical problem.

This does not make drug discovery beyond AI. It means the field should stop borrowing AlphaFold's aura while ignoring the conditions that made AlphaFold possible.

The lesson is not *AI can calculate anything*.

The lesson is *AI becomes powerful when the representation contains the structure of the real problem and the predictions are forced against physical truth*.

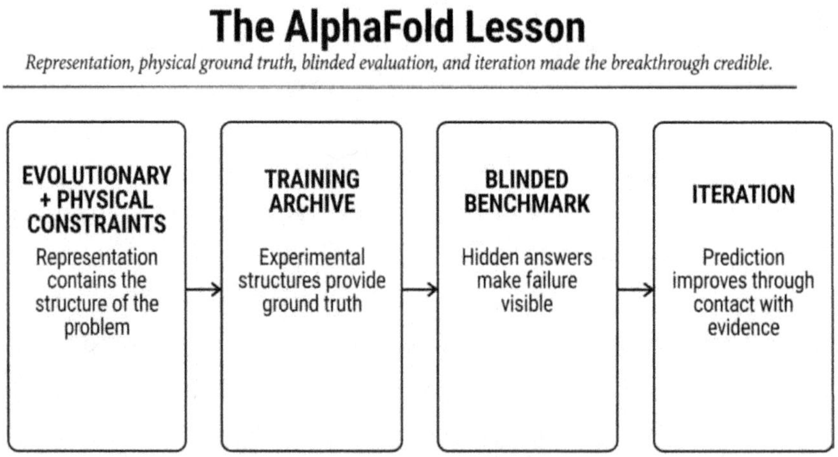

Figure 15. The AlphaFold lesson: representation, ground truth, blinded evaluation, and iteration.

AlphaFold 3 widens the object - and the caution

AlphaFold 3 moved from predicting individual protein structures and selected assemblies toward a unified all-atom representation of biomolecular complexes. Proteins, DNA, RNA, small-molecule ligands, ions, and modified residues could be modeled together. The reported benchmarks showed substantial gains in several interaction classes.

For drug discovery, this matters because proteins do not act alone. Binding pockets can reorganize. Metals and cofactors shape chemistry. Nucleic acids and post-translational modifications alter recognition. Antibodies bind surfaces rather than enzyme pockets. A system capable of co-modeling several molecular types can generate hypotheses that older protein-only workflows could not.

The caution is equally important.

A predicted complex is a static structural hypothesis. It does not directly provide binding affinity, kinetics, residence time, conformational populations, cellular concentration, water networks, membrane context, or the effect of metabolism. Small-molecule pose prediction can be sensitive to chemical representation and training similarity. Stereochemistry and physically implausible clashes require scrutiny. A high-confidence image can still be wrong in the way that matters for a medicinal-chemistry decision.

The model should therefore change the order and quality of experiments, not abolish experiments.

A predicted binding mode can suggest which analog to synthesize, which residue to mutate, which competing mechanism to test, or which target is implausible. Cryo-EM, crystallography, NMR, biophysics, and cellular assays remain the arbiters.

AlphaFold's most useful cultural contribution may be the confidence to ask harder structural questions paired with a clearer understanding that prediction is a hypothesis generator.

What a structure gives - and withholds

A predicted protein structure can transform a program. It can suggest pockets, conformational relationships, mutation effects, complex interfaces, and experiments that would otherwise be difficult to prioritize. AlphaFold 3 and related systems broaden the ambition toward interactions among proteins, nucleic acids, ligands, ions, and modified residues.

But a structure is a state under assumptions.

Proteins move. They occupy ensembles. They are modified, crowded, bound to membranes, exposed to pH and redox gradients, and regulated by partners. A predicted complex may identify a plausible geometry without establishing affinity, kinetics, intracellular exposure, or the biological consequence of binding. A ligand pose does not prove that the ligand reaches the protein in a cell. A beautiful interface does not establish that modulation helps a patient.

The distinction is especially important for natural products and ghost molecules. Large, flexible, or chameleonic structures may present several conformations. Covalent or metal-associated states can change. A molecule may act through induced proximity, membrane perturbation, transport, metabolism, or several targets rather than the single complex selected for visualization.

Structure therefore changes the next experiment rather than ending the investigation.

It can tell a chemist which analog might distinguish two binding hypotheses. It can suggest a mutation that should weaken engagement. It can guide a biophysical assay. It can reveal why a known liability may be structurally linked to efficacy. It can help explain a phenotype after the physical molecule has been confirmed.

The most productive AI systems will be judged by the experiments they make more informative, not by the number of plausible pictures they create.

The dataset is the hidden laboratory

Every drug-discovery model inherits a history of measurement.

Public chemical databases contain millions of structures and bioactivity records. They are indispensable. They are also shaped by decades of choices: which molecules were synthesized; which targets were fashionable; which assays were affordable; which positive results were published; which negative results disappeared; which chemical classes were easy to store; and which property filters removed candidates before data were generated.

A model trained on that record learns chemistry and selection bias together.

This does not make the model useless. It makes provenance part of the model.

If beyond-Rule-of-Five natural products, unstable metabolites, stereochemical families, multi-component effects, local exposures, and context-dependent chemistry are underrepresented, the model will be less reliable there. If most labels describe one target under one assay condition, the model may optimize affinity while missing tissue distribution, network behavior, or toxicity.

The moat is not only the algorithm.

The moat is what the algorithm is allowed to see.

GeneSweep returns here. The genome project generated extraordinary data, yet the cultural model surrounding the data initially expected gene count to explain complexity. Chemical AI can make the same mistake at greater speed: treat the available database as a neutral census of viable chemistry rather than a record of previous institutional choices.

The missing molecules do not appear as blank rows. They are absent from the table.

A Molecular String Is Not an Experiment

A molecular string is a recipe, not a specimen. Open a chemical database and every structure appears with the same visual authority: atoms, bonds, formula, molecular weight, and perhaps a predicted three-dimensional shape. Yet those clean drawings can represent radically different realities. One may be a molecule recovered from a living system, purified, structurally confirmed, and linked to a biological activity. Another may have been synthesized once but never tested. Another may be one example in a patent—or one member of a much larger claimed family that was never individually made. Another may belong to a combinatorial or make-on-demand collection. Another may have been enumerated because it satisfies bonding and stability rules. Another may have been proposed by an artificial-intelligence model and may never have existed outside a computer.

These are not useless or dishonest records. They are different states of evidence. A molecule produced through synthetic biology proves that an engineered living system can make it. A synthesized and assayed compound carries direct experimental evidence under defined conditions. A vendor or patent structure may carry evidence of synthetic accessibility, although that must be verified record by record. An enumerated or generated structure is a hypothesis. When all of them are converted into the same molecular representation and placed in one training table, their provenance can disappear. The model sees a row unless we teach it what kind of reality produced the row.

The scale becomes astonishing almost immediately. One enumerated chemical database generated 166.4 billion plausible structures while stopping at only seventeen heavy atoms—well below much of the chemical territory explored by biological macrocycles and signaling molecules. Expanding molecular size, stereochemistry, ring architecture, and functional groups causes the number of possibilities to grow far faster than any laboratory can synthesize and test them. Biology adds another expansion: the same structure can behave differently by conformation, concentration, tissue, binding partner, metabolism, and time.

This is why computation cannot turn an effectively infinite design universe into validated biology by scoring alone. It can organize, compare, rank, reject, and select the next experiment. It can validate its performance against answers that have already been physically established. But one predictive system cannot confer reality on another predictive system merely because their scores agree. A predicted target, predicted exposure, predicted safety profile, and predicted synthetic route can create an internally consistent story about a molecule that has never touched a cell.

A billion molecular strings are not a billion experiments.

Biological ground truth changes the prior. A physically recovered feature may carry a spectrum, fraction behavior, habitat, time, producing community, spatial distribution, activity pattern, and a record of failed as well as successful tests. None of that makes it a medicine. It makes it a denser question. The future is not nature instead of computation. It is biological reality selecting better starting points, computation interrogating and expanding around them, and experiments deciding which possibilities survive.

The danger of near-limitless generation is not imagination. It is the dilution of evidence.

Generation is not validation

Generative chemistry expands the number of candidate structures. Validation determines whether any candidate works in biology.

The two are often discussed as though they were the same acceleration. They are not.

Computation can validate a model against protected answers; it cannot biologically validate a molecule that has never been made, observed, or tested.

A model can propose a molecule with predicted potency, solubility, selectivity, and synthetic accessibility. The predictions may be useful. The molecule must still be made, structurally confirmed, tested in orthogonal assays, evaluated for exposure and metabolism, challenged against off-targets, dosed in living systems, and ultimately compared with human disease.

The bottleneck moves.

If generation becomes nearly free, experimental capacity and biological relevance become more valuable. The competitive advantage shifts from who can propose the most molecules to who can run the most informative learning loop.

This is a subtle but important reversal. Before generative AI, chemical ideas were expensive. A medicinal chemist had to imagine a structure, assess feasibility, and justify synthesis. When ideas become abundant, the scarce resource is no longer imagination. It is discriminating evidence.

A thousand plausible molecules can be less valuable than one experiment that reveals why the entire design premise is wrong.

Coverage is not the same as scale

Modern foundation models can be pretrained on enormous collections of molecular strings, graphs, structures, papers, patents, images, and biological annotations. The scale is unprecedented.

Coverage can still be narrow.

A dataset containing hundreds of millions of drug-like structures may densely cover one neighborhood of chemistry while leaving natural macrocycles, lipids, metallated states, transient metabolites, reactive species, and community-dependent products sparsely represented. The model can know more examples than any scientist in history and remain uncertain exactly where the next useful biology lies.

This matters because model confidence often reflects familiarity. A structure close to the training distribution can receive coherent predictions across several systems. A novel scaffold may be penalized not because it is biologically implausible but because the record contains few comparable examples.

The result can become a feedback loop. Models prioritize familiar chemistry. Familiar chemistry receives more experiments. The new results strengthen the familiar region of the database. Underrepresented chemistry remains underrepresented.

Throughput improves the map already in hand.

The frontier requires new ground truth.

The danger of synthetic confidence

Generative models can produce structures that look plausible and score well under other models. A target predictor favors the molecule. A property model predicts good exposure. A toxicity model finds no obvious liability. A retrosynthesis system provides a route.

When the systems share similar training histories, errors can reinforce one another. The candidate acquires the appearance of validation without touching a living experiment.

I call this synthetic confidence.

The antidote is experimental friction.

Make the molecule. Confirm its identity and stereochemistry. Measure free concentration. Use orthogonal assays. Test close analogs and inactive controls. Track metabolites. Challenge the model where its training data are sparse. Preserve negative results. Ask whether the prediction survives a tissue, community, animal, and eventually a patient.

Friction is not the enemy of acceleration.

It keeps acceleration connected to reality.

AlphaFold's credibility came from comparison with experimental structures. A drug-discovery model needs an equally honest relationship with synthesis, assay, exposure, safety, and phenotype.

Where AI is already changing discovery

Artificial intelligence is not one technology entering one stage of research. It is spreading across the chain.

Language models extract relationships from literature, clinical records, patents, and internal reports. Image systems classify pathology and cellular phenotypes. Graph and sequence models prioritize targets and predict molecular properties. Generative systems propose chemical structures and proteins. Retrosynthesis tools suggest routes. Robotic laboratories execute experiments and feed results back into active-learning systems.

Each application has a different standard of success.

Literature synthesis is valuable when it retrieves the right evidence and preserves uncertainty; it is dangerous when fluent text hides invented sources. Property prediction is valuable within a defined domain; it is dangerous when applied to novel chemistry without calibrated uncertainty. Generative design is valuable when the molecules are made and tested; it is dangerous when model scores are reported as biological accomplishment.

The discovery chain also contains asymmetry. An error in prioritizing one of ten thousand papers may waste time. An error in a toxicology prediction can advance a dangerous compound. The closer a model moves toward patient decisions, the greater the need for validation, governance, and interpretable evidence.

The best systems will not ask AI to replace scientists. They will redesign the division of labor. Machines can organize enormous search spaces, recognize patterns, and maintain complex experimental histories. Humans can define meaningful questions, detect representation errors, integrate tacit knowledge, and make accountable choices under uncertainty.

The competitive advantage will come from the loop, not the model in isolation.

AI's best role in the ghost-genome problem

The strongest near-term roles for AI in hidden chemistry are not unconstrained invention. They are translation and prioritization.

Spectral interpretation. Models can infer formulas, substructures, fragmentation behavior, and candidate classes from tandem mass spectra.

Molecular-family mapping. Embeddings can organize millions of unannotated spectra and identify recurring chemical neighborhoods before every member is named.

Genome-metabolome linking. Models can rank gene clusters or enzyme families likely to produce observed molecular classes while preserving uncertainty.

Target and phenotype prediction. Multi-task systems can prioritize receptor families, pathways, and assays, especially when trained on broader chemical space and negative data.

Analog design. Once a physical scaffold is known, generative methods can explore nearby structures that preserve key geometry while changing exposure, potency, stability, or liability.

Experiment selection. Active learning can choose the next synthesis, isotope label, perturbation, or assay expected to reduce uncertainty most efficiently.

Literature integration. Language models can connect fragmented findings across chemical ecology, pharmacology, clinical medicine, and patents - provided the underlying sources are checked.

The common theme is constraint.

AI is most useful when the question contains physical evidence.

The economics of the next experiment

As candidate generation becomes cheaper, the scarce resource changes.

The constraint may be synthesis slots, purified protein, primary human cells, animal capacity, analytical time, medicinal-chemistry attention, or the number of biological questions a team can interpret without losing coherence. In every case, the value of AI shifts from producing possibilities to allocating experiments.

This is an economic problem as much as a computational one.

Suppose a wet-lab experiment costs ten thousand dollars and two weeks. A model that improves the probability of an informative result from ten percent to twenty percent has created enormous value even if it does not directly nominate a clinical candidate. A model that generates one million attractive structures but leaves the hit rate unchanged may create cost by increasing triage.

The correct objective is often expected information gain per unit of constrained resource.

An experiment can be valuable even when the result is negative if it separates hypotheses. A close analog that loses activity can identify a necessary functional group. An assay selected to distinguish membrane disruption from receptor binding can eliminate an entire mechanistic branch. A spatial sample can reveal that a molecule never reaches the proposed site. A synthesis failure can expose an unrealistic design region.

Active learning formalizes part of this logic. The model selects the next candidate expected to reduce uncertainty or improve the objective. But the choice of uncertainty matters. A model can become very certain about a narrow assay while the clinically decisive uncertainty remains exposure or safety.

Human judgment must define the decision the experiment is meant to improve.

The closed-loop laboratory of the future will therefore include an explicit hierarchy:

- scientific uncertainty: what mechanism or structure is true?
- development uncertainty: can the molecule be made, delivered, and tolerated?
- portfolio uncertainty: is this the best use of the next dollar and month?

- strategic uncertainty: does the result validate a reusable capability or only one asset?

AI can help quantify and update those uncertainties. It cannot decide their value without the purpose supplied by people and patients.

The base-rate problem in plain language

Suppose a screening system correctly recognizes ninety percent of viable candidates and falsely labels only one percent of nonviable candidates as promising. Those numbers sound excellent.

If one candidate in ten is truly viable, the positive set will contain many real opportunities.

If one candidate in a million is viable, a one-percent false-positive rate produces roughly ten thousand false positives for every true one. The screen remains statistically accurate by one measure and operationally useless by another.

Positive predictive value depends on prevalence.

Drug discovery does not have one universal prevalence because viability depends on indication, modality, stage, and definition. The thought experiment shows why expanding chemical space can create a hidden penalty. If the search universe grows much faster than the subset satisfying efficacy, exposure, safety, manufacturability, and clinical utility, the base rate falls.

AI can help in two ways. It can reduce false positives by modeling more constraints. It can increase prevalence by starting from a better distribution.

Most public attention focuses on the first. The ghost-genome argument focuses on the second.

A natural scaffold that has existed in a living system does not solve safety or efficacy. It may raise the chance that the structure can be made, remain coherent in biological media, and interact with living matter in an organized way. Habitat, biosynthetic family, and phenotypic activity can add further information.

The prior becomes stronger as context becomes richer.

A random virtual molecule has a formula and predicted properties. A recovered natural feature can have a formula hypothesis, spectrum, fraction behavior, producing community, spatial distribution, activity pattern, and evolutionary neighborhood.

The comparison is not imagination versus nature. It is a sparse prior versus a dense prior.

Finite natural priors

In my work on Eroom's Law, I have proposed a formal intuition: if the searchable design universe expands faster than the prevalence of compounds satisfying all the requirements of a viable drug, then positive predictive value can decline even as screens become more sensitive.

A viable drug must satisfy efficacy, safety, exposure, selectivity, manufacturing, formulation, and clinical utility. If the number of imaginable structures grows toward the effectively infinite while the viable subset grows more slowly, the base rate falls.

This is prevalence dilution.

A physically recovered natural molecule begins in a finite, realized system. It is not necessarily safe or drug-like. It may be a toxin, an ecological weapon, or a useless curiosity. But biology made it. That fact raises the prior probability of manufacturability by living chemistry, structural stability under some conditions, and participation in a biological interaction.

I call these starting points finite natural priors.

The proposed strategy is hybrid. Natural-product access enriches the starting distribution with biologically realized structures. Synthetic chemistry and AI then explore neighborhoods around those structures, optimize liabilities, infer targets, design analogs, and select experiments.

The hypothesis is not *nature beats AI.*

It is *AI performs better when biological ground truth improves its prior.*

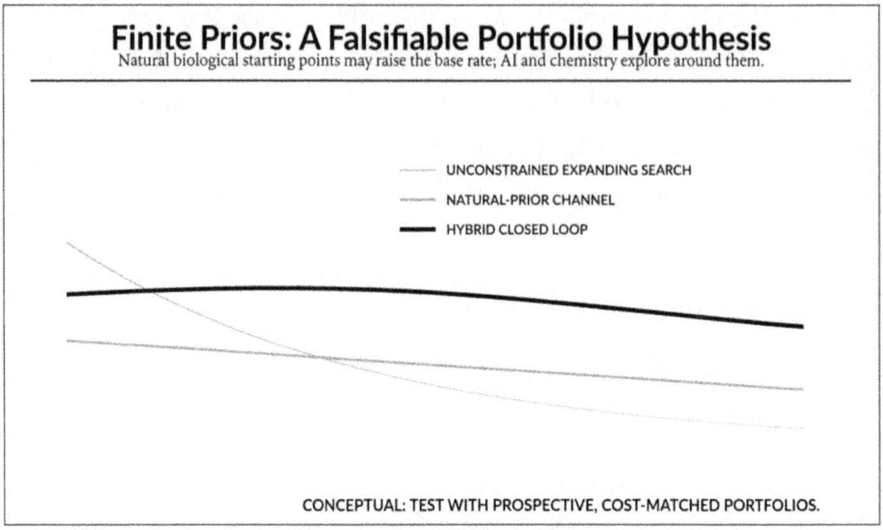

Figure 16. Finite priors: a falsifiable portfolio hypothesis comparing unconstrained design, natural-prior discovery, and a hybrid closed loop.

How to test the claim

The finite-prior argument should not be validated by listing famous natural-product medicines after the fact. It requires a prospective comparison.

Matched discovery portfolios could be established across three channels:

1. unconstrained synthetic or generative design;
2. natural-prior discovery and optimization;
3. a hybrid channel using natural scaffolds, spectra, or mechanisms to guide design.

Programs would be matched for disease difficulty, budget, assay quality, and stage. Endpoints would include confirmatory hit rate, false-positive rate, novelty, developability, time to validated lead, cost per lead, and later attrition.

The model predicts that the hybrid will outperform when natural priors meaningfully increase the prevalence of viable structures and when the cost of access, isolation, resupply, and optimization does not erase the advantage.

It could fail.

Natural-product isolation may be too slow. Ecological functions may be irrelevant to human disease. Supply may remain difficult. Generative systems may learn enough physics and biology to improve base rates without natural starting points. The hybrid may win in some target classes and lose in others.

Good.

Those outcomes can be measured.

The purpose of a theory is not to protect a worldview. It is to make experiments and capital more intelligent.

AI can imagine chemistry almost without limit. Biology can validate only finitely. The strategic advantage belongs to those who begin with better priors.

Aggressive About the Question, Conservative About the Claim

The phrase breakthrough aggressiveness can sound like a license to move fast and believe hard. I mean something more disciplined.

Be aggressive about whether the problem has been framed correctly. Be willing to redesign the instrument, cross disciplines, and pursue an anomaly that ordinary workflows erase. Challenge the assumption that more throughput inside the inherited architecture is the only form of progress.

Then be conservative about the claim. Register predictions before results are known. Keep failed molecules in the denominator. Use independent standards. Distinguish a model score from a measured affinity, an affinity from a cellular mechanism, and a mechanism from patient benefit.

This asymmetry is essential. Conservatism at the level of the question funds yesterday. Aggressiveness at the level of the conclusion creates hype. The useful combination is a radical question followed by an unforgiving experiment.

AlphaFold embodied that combination. The ambition was enormous. The benchmark was blind. The answer was physical. Drug discovery needs the same willingness to be embarrassed by reality.

The global race will be won by learning systems

The distinction between frontier and factory no longer maps cleanly onto countries.

For years, Western biotechnology often described China as a lower-cost execution base: synthesis, screening, and manufacturing performed quickly after ideas originated elsewhere. That description is increasingly obsolete. Chinese companies and research institutions now originate globally competitive assets, develop novel modalities, and license innovation into multinational pipelines. The speed and integration of the experimental factory can feed the frontier.

The United States retains major advantages in university science, venture formation, capital markets, immigration, and cultural permission to form new companies around uncertain ideas. Europe, Japan, South Korea, Singapore, India, and other ecosystems contribute distinctive capabilities.

The strategic question is not which flag owns innovation. It is which organizations connect exploration, physical data, computation, and execution with the least friction.

A factory without a frontier optimizes yesterday. A frontier without a factory produces papers and prototypes. A learning system joins them.

For the ghost genome, the winning platform must collect and preserve unusual biology, measure it rigorously, use AI to prioritize uncertainty, synthesize and test candidates quickly, and transfer validated assets into development. The loop must cross disciplines and often organizations.

No specialist owns the whole loop. The loop itself is the instrument.

The countries and companies that treat data provenance, negative experiments, and physical samples as strategic infrastructure will have an advantage over those that treat AI as a software layer added to the same old candidate distribution.

The material loop

Industrial-scale access matters in this framework because it produces a different kind of training set.

A useful record links a physical feature, high-resolution spectra, fraction behavior, habitat, time, genomic context where available, biological activity, structural confidence, synthesis status, and negative as well as positive results.

This is biological ground truth: a traceable connection between material, measurement, context, and function.

Competitors can buy compute. They can reproduce an algorithm described in a paper. They cannot instantly reproduce the ecological history, physical samples, fractionation choices, assay failures, and tacit knowledge accumulated through years of access.

The data advantage is material before it is digital.

The closed loop is the real discovery engine:

1. physical biology produces molecules and spectra;
2. AI organizes the unknowns and proposes the most informative next step;
3. chemistry and biology test the proposal;
4. the results - especially the failures - update the model;
5. the updated model changes what is collected, isolated, synthesized, or assayed next.

AlphaFold taught the world what happens when data, representation, constraints, and ground truth align.

The ghost-genome challenge is to build that alignment for a chemical world the current databases only partly contain.

A model that knows where it is ignorant

Confidence is useful only when it is calibrated.

An AI system operating inside well-represented chemical space should be able to distinguish that familiar territory from a macrocycle, spectral family, or community state far outside its training distribution. The system should not respond to unfamiliarity by producing the same style of score with unjustified precision.

For ghost chemistry, uncertainty should attach to several layers:

- Is the feature real and reproducible?
- Is the molecular formula constrained?
- Is the connectivity known?
- Is stereochemistry resolved?
- Is the proposed source supported?
- Is the activity direct, indirect, or mixture-dependent?
- Is the target prediction inside the model's validated domain?
- Is exposure measured in the relevant compartment?

A strong platform exposes these layers to the user. It can say, in effect: *I am confident that these spectra form a family; moderately confident that the scaffold contains this substructure; weakly confident about the complete structure; and unable to extrapolate the permeability model to this conformational class.*

That answer is more valuable than one synthetic probability attached to the entire story.

The architecture also improves governance. Teams can decide which uncertainty is worth reducing and which is acceptable at the present stage. A discovery hypothesis may advance with partial structural knowledge if the fraction is reproducible and the next experiment is designed to resolve identity. A development candidate cannot.

Knowing where the model is ignorant is not a concession.

It is the condition for learning.

The laboratory of the near future

A closed-loop natural-products laboratory could begin the day with millions of unresolved spectra rather than millions of imagined structures.

The system groups molecular families, compares habitat and activity, and identifies a cluster of features associated with a reproducible phenotype. It proposes several structural hypotheses and selects the fractionation step most likely to distinguish them. A robotic system performs microseparations. Spectra update. One hypothesis falls. Another gains confidence.

The model then chooses between two expensive actions: synthesize a candidate standard or collect additional source material under a controlled environmental condition. It estimates which experiment will reduce uncertainty most. The chosen experiment is performed. The standard matches the spectrum but not the biological activity, revealing that the active feature is a coeluting analog or mixture.

That information: success, nothing, failure becomes training data.

A new separation isolates the analog. Cellular assays show a coherent phenotype. Chemoproteomics identifies several engaged proteins. Analog design explores the natural scaffold's neighborhood. Exposure and toxicity models prioritize compounds, but every prediction is tied to synthesis and measurement.

This laboratory is not science fiction in its individual components. The challenge is integration, data quality, and economics.

The intelligence comes from the loop's memory. It remembers not only what worked but what failed, under which conditions, and with which physical material.

The future AI scientist will not float above the laboratory.

It will be embedded in the chain of custody between ecology and evidence.

The final story follows a molecule through the complete chain: soil, organism, screen, chemistry, manufacturing, regulation, partnership, and global delivery. It shows why discovery is only the beginning - and why one physical natural prior can matter to hundreds of millions of people.

CHAPTER 10

The Soil Microbe That Changed Global Health

From Natural Prior to Medicine

STORY 10

THE SOIL MICROBE THAT CHANGED GLOBAL HEALTH

From Natural Prior to Medicine

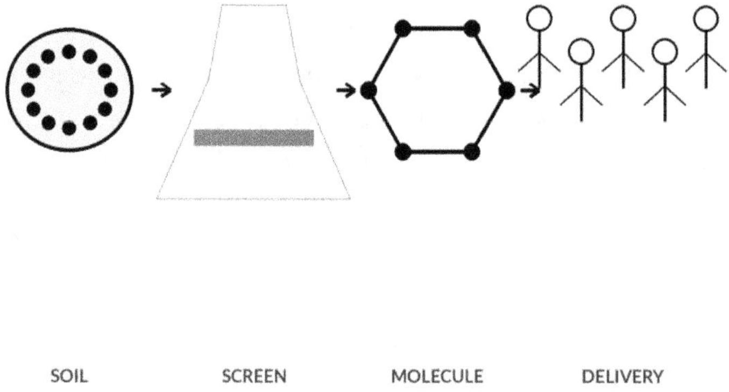

THE TURNING POINT

Discovery, development, delivery, and stewardship completed the chain.

Satoshi Ōmura did not search soil by admiring its complexity. He built a disciplined system for turning complexity into candidates.

In the late 1960s and early 1970s, the Japanese microbiologist and bioorganic chemist collected thousands of soil samples, cultured microorganisms, and screened their products for useful biological activity. Soil had already yielded antibiotics and other medicines. The challenge was not believing that nature contained interesting chemistry. It was recovering organisms, maintaining them, recognizing unusual activity, and choosing which few samples deserved expensive follow-up.

A collaboration with Merck connected Ōmura's collection and screening work at the Kitasato Institute to industrial pharmacology in New Jersey. Promising cultures were sent to Merck Research Laboratories, where William Campbell led testing against parasitic worms.

The workflow was simple enough to describe and demanding enough to execute. Mice were infected with nematodes. Culture samples were administered. Investigators looked for animals cleared of parasites without unacceptable harm.

One culture was extraordinary.

It came from a *Streptomyces* strain isolated from soil collected near a golf course in Kawana, southwest of Tokyo. Merck scientists isolated a family of related compounds and named them avermectins, reflecting their ability to make infected animals free of worms.

The natural product was not the finished medicine.

Chemists separated the family, compared activity, and synthesized large numbers of derivatives. A modest chemical change - hydrogenation at a particular bond - produced ivermectin, a mixture of closely related compounds with improved potency and pharmaceutical properties.

The drug became a remarkable veterinary product against worms, mites, ticks, and other parasites. Commercial success created manufacturing knowledge, safety data, and confidence. Campbell and colleagues then asked whether the compound could address human filarial diseases, including onchocerciasis - river blindness.

River blindness is transmitted by black flies breeding near fast-flowing water. The parasitic worms can live for years. Their offspring move through skin and eyes, producing intense itching, disfiguring disease, and blindness. The communities most affected were often remote and poor, with little ability to support a conventional commercial market.

Clinical development required additional champions, including Merck physician Mohammed Aziz, and collaboration with international health organizations. Ivermectin could dramatically reduce the microfilariae responsible for symptoms and transmission. But a medicine in a factory does not treat anyone until it reaches the patient repeatedly, safely, and at enormous geographic scale.

In 1987, Merck made a commitment that changed the history of pharmaceutical philanthropy: donate Mectizan, the human formulation of ivermectin, as much as needed and for as long as needed for river blindness.

The promise created a second innovation challenge. Tablets had to reach remote communities. Distribution could not depend on health systems that did not exist. The Mectizan Donation Program brought together Merck, the World Health Organization, the Task Force for Global Health, ministries, researchers, nongovernmental organizations, and local communities. Community-directed treatment models trained local distributors and made repeated mass administration possible.

Over the following decades, billions of treatments were donated. Several countries eliminated transmission. The program expanded to lymphatic filariasis in settings where Mectizan could be combined with donated albendazole.

In 2015, Ōmura and Campbell shared the Nobel Prize in Physiology or Medicine with Youyou Tu.

The complete chain is what makes the story the right ending for this book:

environmental microorganism -> physical molecule -> biological activity -> medicinal chemistry -> manufacturing -> clinical evidence -> global delivery -> public-health impact

Every arrow required a different institution, discipline, incentive, and form of trust.

The soil sample did not become a medicine because nature was wise.

It became a medicine because frontier discovery and factory execution eventually became one system.

Discovery was a partnership before the molecule had a name

The Kitasato-Merck arrangement is an early example of the frontier-and-factory model working by design.

Ōmura's group possessed access, microbiological skill, and a disciplined method for collecting and triaging environmental organisms. Merck possessed disease models, medicinal chemistry, pharmacology, toxicology, manufacturing, and development infrastructure. The agreement allowed promising cultures to move across an institutional boundary without requiring either side to become the other.

The soil strain was valuable because the partnership preserved several forms of optionality. The organism could be fermented. The product family could be isolated. Related compounds could be compared. Chemistry could modify the scaffold. Veterinary indications could establish utility and economics before human use.

Avermectin was a family, not one perfect molecule. Medicinal chemistry did not reject the natural product's complexity; it learned which portion of the family carried the best balance and then made a targeted change. The hydrogenated derivatives that became ivermectin preserved the architecture that evolution had built while improving the product for human purposes.

This is what *natural prior plus design* looks like in practice.

The veterinary market also mattered strategically. Parasite control in livestock and companion animals created a large commercial product, manufacturing scale, field experience, and safety knowledge. Human river blindness affected communities with little purchasing power. Without the earlier value chain, the later donation would have been more difficult to imagine and supply.

The sequence is uncomfortable for simple narratives about profit and purpose. Commercial success in one market helped enable public-health delivery in another. The same company that developed a highly valuable product also made an open-ended donation commitment because the ordinary market could not reach the people who needed it.

Innovation systems are rarely morally pure. They can still be designed to produce extraordinary public value.

From culture flask to human medicine

The distance between an active fermentation product and a medicine is easy to compress in hindsight.

After the Kitasato-Merck collaboration identified the avermectin family, researchers had to determine which components carried the most useful activity, how to improve the profile, how to manufacture material consistently, and how to characterize safety across species and doses. Chemists modified the natural products to create ivermectin, a derivative with a favorable antiparasitic profile.

The molecule first proved its value in animal health. That route provided pharmacology, manufacturing experience, and commercial evidence while the human opportunity developed. The transition to human disease required different studies, dosing logic, safety expectations, and public-health partnerships.

River blindness is caused by *Onchocerca volvulus*, a parasitic worm transmitted by blackflies near fast-flowing rivers. Adult worms can survive for years and produce microfilariae that migrate through skin and eyes, provoking inflammation, severe itching, skin disease, and blindness. The geography that gives the disease its name also makes delivery difficult. Endemic communities can be remote, health infrastructure limited, and repeated treatment necessary.

A medicine effective in a controlled trial would not be enough. The product had to be manufactured at scale, donated or financed, distributed through programs that reached villages repeatedly, monitored for safety, and integrated with local trust and public-health knowledge.

The discovery chain therefore continued after regulatory evidence.

A molecular scaffold became a drug. The drug became a program. The program became a durable network of people, logistics, training, surveillance, and commitment.

This is why the final translational unit is not always the pill.

It is the system capable of placing the pill in the right hands for as long as the disease requires.

Delivery became the second discovery

The biology of ivermectin was not the final bottleneck. Distribution was.

River-blindness-endemic communities were often located far from roads, clinics, laboratories, and reliable health records. Treatment had to be repeated. Dosing needed community trust and local knowledge. Programs had to identify who should receive medicine, monitor safety, maintain supply, and reach populations across countries and conflict zones.

The Mectizan Donation Program created governance around the promise. International agencies, ministries, nongovernmental organizations, researchers, and community distributors had to coordinate. A tablet manufactured to pharmaceutical standards entered a delivery system that depended on people walking it the last miles.

Community-directed treatment was itself an innovation. Communities helped decide how distribution would occur and selected local distributors. The model strengthened local ownership and made mass treatment more sustainable than a purely external campaign.

The lesson for the ghost genome is that translation includes the system around the molecule.

A remarkable drug can fail through manufacturing cost, cold-chain requirements, diagnostic complexity, reimbursement, adherence, or geography. A modest medicine can transform public health when product and delivery architecture fit the setting.

The first lighthouse asset should therefore be selected not only for biological potency but for the plausibility of the complete path to a patient.

The promise made in 1987

In 1987, Merck announced that it would donate Mectizan for river blindness for as long as needed.

The sentence changed the development problem.

A conventional commercial model could not easily reach populations with great medical need and limited ability to pay. Without a delivery commitment, a scientifically successful medicine might have remained a demonstration of what was possible rather than a public-health transformation.

The donation program did not operate alone. International agencies, ministries of health, nongovernmental organizations, local health systems, researchers, and community-directed distributors became part of the architecture. In many places, people from the affected communities organized treatment delivery themselves. Over time, the program expanded to lymphatic filariasis in appropriate settings and contributed to elimination goals across multiple countries.

The story contains a difficult lesson for innovation narratives. A company can be essential without being the only hero. A natural product can be remarkable without delivering itself. A donation can be generous and still depend on public infrastructure, local labor, scientific surveillance, and sustained political commitment.

The result was cumulative. One annual treatment did not erase transmission everywhere. Programs had to return. Coverage had to remain high. Adverse events and co-endemic infections required attention. Drug supply had to arrive reliably. Progress was measured at population scale.

This is relevant to the future of microbiome-derived therapeutics because access should be designed before success makes it urgent.

A locally delivered therapy may require cold chain, trained administration, diagnostic selection, or repeated monitoring. A highly individualized product may be scientifically elegant and inaccessible at scale. A complex fermentation process may create cost or supply vulnerability. A diagnostic panel may widen inequity if only wealthy systems can act on the result.

The product profile includes the delivery architecture.

Ivermectin became globally consequential because pharmacology and access eventually formed one program.

The frontier and the factory

A large pharmaceutical company is not simply a group of scientists. It is an integrated machine built around asset progression.

Targets are selected. Assays are standardized. Chemistry creates and optimizes series. Toxicology, formulation, manufacturing, clinical operations, regulatory affairs, market access, and commercial teams take over at defined gates. The organization is designed to reduce uncertainty as an asset moves forward.

That machine is one of humanity's great achievements. It can turn a molecular observation into a globally manufactured medicine with extraordinary quality and regulatory control.

The machine also has preferences.

It prefers a molecule that fits known analytical methods. It prefers a mechanism that can be explained to a governance committee. It prefers a biomarker that demonstrates target engagement. It prefers a manufacturing route with precedent. It prefers an indication with a trial design, regulatory path, payer logic, and market.

A platform proposing unknown, context-dependent, multi-target molecules from uncultured communities collides with every preference at once.

The molecule may be difficult to resupply. The producer may be unknown. The mechanism may be a network rather than one target. The relevant assay may require host and microbial components. The intellectual-property strategy may depend on composition, analogs, use, process, and data rather than one clean invention.

From inside the incumbent machine, the platform looks messy.

From outside, the mess may be the unmet biology.

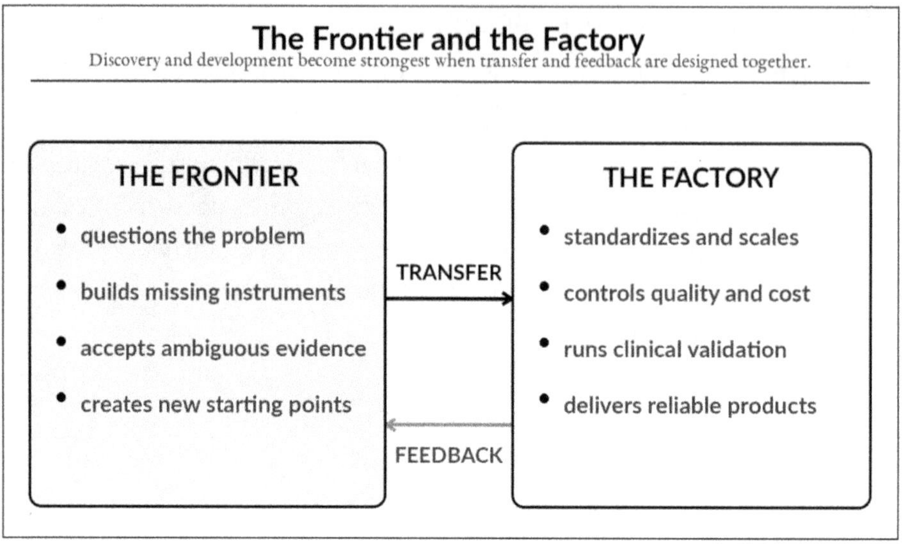

Figure 17. The frontier and the factory: discovery questions the problem; development standardizes, validates, and delivers.

The answer is not to ask the factory to become the frontier in every respect. It is to build a transfer point.

Small companies and academic groups take concentrated scientific risk. Large companies provide capital, clinical infrastructure, manufacturing, regulatory expertise, and global reach. The relationship works when each side respects the other's comparative advantage and when the handoff occurs after the discovery has become sufficiently defined.

The frontier company must be able to access chemistry others cannot access; preserve context; tolerate periods of unknown identity; connect phenotype, spectrum, structure, and exposure; protect the work; kill weak hypotheses; and convert at least one strange observation into a development package.

The development partner must reproduce and manufacture the molecule, establish pharmacology and toxicology, define regulatory strategy, select indications and biomarkers, run trials, and scale supply.

A platform can remain fascinating for years without producing a partner-ready asset. An asset can be handed off too early and fail because the platform has not solved identity, resupply, mechanism, or exposure.

The first commercial objective is not to prove the entire ghost genome.

It is to produce one unambiguous result the existing system cannot ignore.

The first lighthouse

Every platform needs a lighthouse asset - a program that illuminates the broader coast.

The ideal first asset has a large unmet need, a measurable near-term phenotype, a tractable route to synthesis or manufacturing, a safety strategy grounded in exposure, and enough mechanistic evidence to survive diligence. It should have a differentiated reason the starting molecule could not easily have emerged from a standard library.

The largest market is not always the best first proof.

A gut-restricted molecule can be attractive because local action may simplify systemic safety and leverage the site of microbial signaling. A skin or wound application can allow local delivery and visible endpoints. An anti-infective can provide a direct phenotype but face difficult economics. A metabolic program may attract partners while entering an intensely competitive field. A neurorepair program can be transformative and biologically complex.

The platform thesis and the first business thesis are not identical.

Rapamycin's initial antifungal activity did not predict the full mTOR story. Ivermectin's veterinary success helped create the manufacturing and safety foundation for a global human-health program. The first lighthouse may validate access and translation through an indication narrower than the platform's ultimate reach.

The innovator's dilemma in medicine

Clayton Christensen's most unsettling insight was that strong organizations can miss a new architecture because they make good decisions according to the metrics that built the old one.

They listen to important customers. They improve high-margin products. They allocate capital to opportunities with measurable demand. They standardize processes and reduce error. These behaviors create excellence within an existing system.

Then a new system arrives with inferior early performance on the old metrics.

Early digital cameras were worse than film in important ways. Early personal computers looked trivial beside mainframes. Early electric vehicles were expensive, short-ranged, slow to charge, and dependent on infrastructure that barely existed. **Incumbents correctly identified the weaknesses and incorrectly assumed the weaknesses defined the future.**

The electric-vehicle analogy is useful only at the architectural level. An electric vehicle is not merely a conventional car with one component replaced. Battery, software, power electronics, packaging, manufacturing, charging, and service evolve together.

Microbiome-derived chemistry may require an analogous change in discovery architecture. If the source is an intact community, access and preservation become core capabilities. If most chemistry is unknown, measurement and annotation become the library. If function is polypharmacological, phenotypes and networks become first-class. If action is local, systemic exposure is not always the objective. If a product arises from distributed metabolism, a single gene cluster is not the complete map.

A conventional screening group can adopt each tool. It may struggle to reorganize around the whole because its gates were designed for a target and a compound series.

This is why separate units, partnerships, options, and platform companies matter. The new architecture needs enough protection from the old metrics to produce an asset the old system can then evaluate.

Disruption does not mean destroying pharma.

It means changing what pharma treats as the beginning of discovery.

Tuning states rather than removing objects

Medicine has often been organized around removal.

Kill the pathogen. Block the receptor. Suppress the cytokine. Destroy the tumor. Clear the plaque. Lower the concentration. Remove the diseased tissue.

Removal is sometimes exactly right. A bacterial infection can kill. A malignant clone can expand. An overactive immune pathway can destroy an organ.

Many chronic diseases, however, are persistent states. They involve feedback among tissues, immune cells, metabolism, microbes, nerves, behavior, and environment.

The future described in this book is a medicine of state change: shifting the bioreactor from a disease-supporting pattern toward a resilient one.

The intervention may still be one defined molecule. It may influence several related targets at modest exposure. It may remain local. It may change microbial competition rather than act directly on a human protein. It may serve as a diagnostic signal rather than a drug. It may teach medicinal chemists a mechanism that leads to a simpler synthetic analog.

Complex biology does not excuse vague products.

A regulator does not require a drug to have one target. Many approved medicines have multiple mechanisms. Regulators require a defined product, controlled manufacturing, rational dosing, monitored risk, and evidence of benefit for a defined use.

Restores balance is not an endpoint.

Postbiotics before probiotics

The public microbiome market first became familiar through live organisms: probiotics intended to add beneficial bacteria. Live biotherapeutics remain scientifically promising, especially when ecological restoration is the goal. They also face challenges in engraftment, consistency, manufacturing, patient variability, and mechanism.

A chemical approach asks whether the function can be delivered without asking a living organism to establish itself.

The terminology is imperfect. A consensus definition of *postbiotic* emphasizes preparations of inanimate microorganisms and their components that confer health benefit, while commercial usage often extends the word to purified metabolites. The product should be defined rather than hidden behind the label.

For ghost-molecule discovery, the core objective is to **identify the chemical instruction**, choose the best way to reproduce, and deliver it.

The instruction may become an unmodified natural product, semisynthetic analog, fully synthetic mimic, fermentation product, defined mixture, diagnostic biomarker, or molecule that alters microbial ecology rather than human cells directly.

A live organism can provide responsive production in place. A defined molecule provides dose, identity, and manufacturing control. Neither is universally superior.

The clinical question should choose the modality.

Geofencing the molecule not the message

One of the most practical lessons of the human bioreactor is that a drug does not always need to travel everywhere.

A gut-restricted molecule can act on luminal microbes, mucus, epithelium, local immune cells, or enteroendocrine pathways while minimizing systemic exposure. A skin-restricted molecule can influence repair without burdening liver or brain. An inhaled medicine can concentrate in lung. A hydrogel can retain chemistry at a wound or nerve interface.

I use *geofencing* for this deliberate control of biological geography.

The gut is the clearest example because it is not one target or even one organ. It is an information network. A molecule may remain in the lumen, mucus, or mucosa and never reach meaningful concentrations in systemic blood or brain. Yet it can still alter distant physiology. It may change microbial competition or metabolite production. It may engage epithelial, immune, or enteroendocrine cells. Those cells can release hormones, alter local chemistry, or signal the enteric nervous system and the vagus nerve. Downstream metabolites can enter portal blood and reach the liver. The physical molecule may stay inside the gut while the information it initiates travels much farther.

This is the practical meaning of the gut–brain axis. The phrase can sound like a pipe running from bacteria to the brain. It is better understood as a layered, bidirectional network. In systems language, a *node* is simply a place where information is sensed, transformed, relayed, or integrated.

The first-contact nodes include microbes, phages, mucus, epithelial cells, local immune cells, nutrient sensors, and enteroendocrine cells. Relay nodes include gut hormones, the enteric nervous system, vagal sensory fibers, portal circulation, and metabolic organs such as the liver and pancreas. Central integration occurs across circuits governing satiety, hunger, energy expenditure, reward, stress, memory, and learned eating. Different diseases—and different people—may rely on different paths through this network.

The map is still expanding. Known microbial and host molecules connect to known receptors and pathways, but newly recovered molecules may reveal natural ligands for familiar targets, previously unrecognized targets, new cell-state dependencies, or combinations of nodes that current diagrams do not include. Ghost-molecule discovery can therefore expand both sides of the map: it can add new chemical keys and help identify biological locks, relays, and circuits that were previously missing.

The aim is not to strike every node. More targets do not automatically create better pharmacology. The objective is to identify a coherent subset whose combined movement produces a useful physiological state at a tolerable exposure.

For drug design, this creates two separate questions:

Where must the molecule physically go?

Which downstream nodes must the resulting signal ultimately influence?

A gut-retained compound might act first on microbial ecology, barrier cells, immune cells, enteroendocrine signaling, or enteric and vagal interfaces, then propagate an effect toward satiety, hunger, energy regulation, or reward systems without the compound itself entering the brain. That chain must be demonstrated through measured relays—local exposure, proximal target or cellular activity, hormonal or neural signals, and a relevant phenotype—not assumed from a computational target map.

Geofencing can make polypharmacology safer. A molecule with useful activity across several gut-facing nodes may be unacceptable if the same free concentration reaches heart or brain. Reducing permeability, increasing polarity, using a local formulation, encouraging efflux, or creating rapid first-pass metabolism can turn an apparent liability into a therapeutic boundary.

The reverse can also be engineered when systemic or central action is required. The same scaffold family may support local and systemic versions, each designed for a different exposure map.

The mucus phages return here. Their effect depended on concentration at a boundary. The twins return as well. Their metabolic phenotype depended on community, diet, and communication with the host—not merely on the presence of one organism.

Location and context are not secondary properties added after target selection.

They are design variables.

The molecule may stay behind. The message can travel.

Three possible therapeutic geographies

The future can be made more concrete by examining where a molecule might act.

The gut. A gut-restricted molecule could enter the gut–brain and gut–metabolic network at one or more local nodes: microbial ecology, mucus, epithelial repair, local immunity, enteroendocrine cells, or enteric and vagal interfaces. In metabolic disease, the objective might be a measured pattern across meal initiation, satiety, motility, barrier function, inflammation, energy regulation, and reward—without requiring the molecule itself to enter the brain. It would not need to replace incretin therapies to be valuable. It could provide another chord, route, maintenance strategy, or complementary mechanism for a defined patient group.

The wound or nerve interface. Injury creates a temporary ecosystem of damaged tissue, inflammatory cells, matrix, microbes, oxygen gradients, and pain signaling. A locally retained molecule could influence inflammation, fibrosis, vascularization, infection, and regeneration over time. The delivery material - hydrogel, wrap, scaffold, or coating - becomes part of the pharmacology.

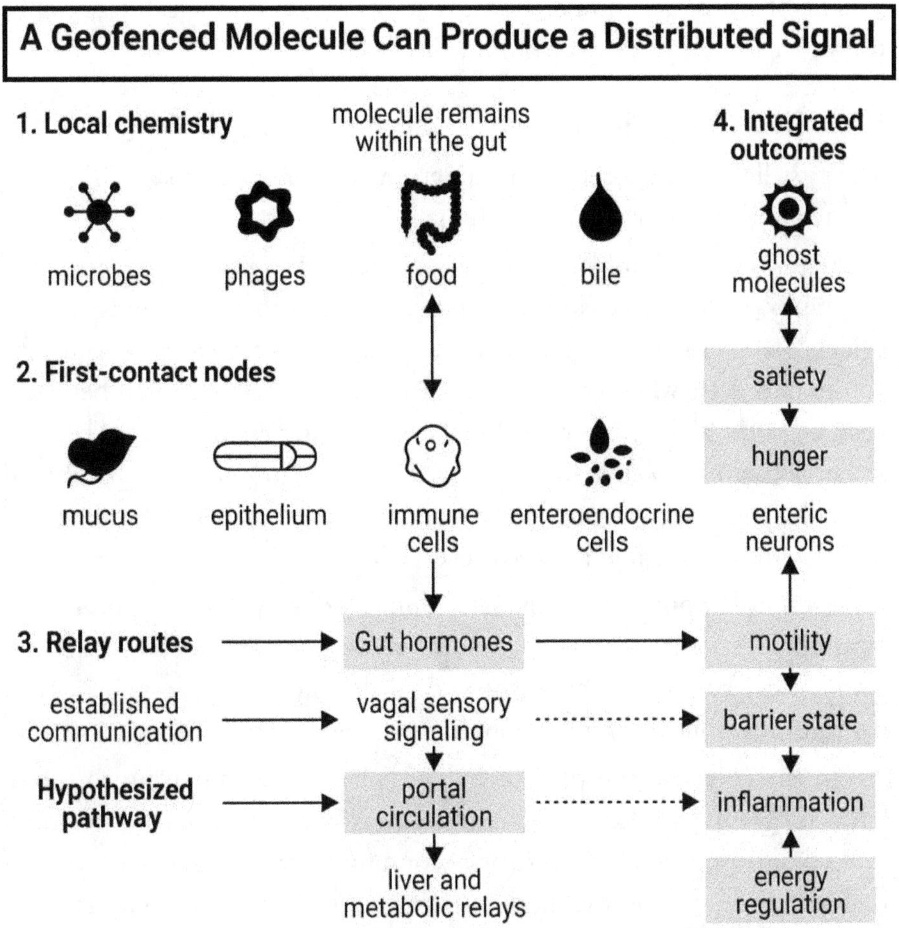

Figure 18. A geofenced molecule can produce a distributed signal. A gut-restricted molecule may remain in the lumen or mucosa while influencing distant physiology through microbial, epithelial, immune, enteroendocrine, enteric-neural, vagal, hormonal, and metabolic relays. Dashed paths represent emerging or hypothesis-level relationships.

The skin or oral cavity. These surfaces allow direct access, visible or measurable endpoints, and strong exposure control. A molecule can act on barrier repair, biofilm state, inflammation, or local immunity while limiting systemic burden.

These examples are research architectures, not claims of efficacy for any undisclosed program. They illustrate why geography can make complex chemistry developable.

The first product should choose a site where the biological state can be measured and the risk can be bounded.

Lighthouse candidates and the evidence package

The first lighthouse asset from hidden microbiome chemistry should be selected for how clearly it can teach the field.

A gut-restricted molecule might be attractive because the relevant geography is accessible and systemic exposure can be minimized. The evidence package would still need to show where within the gut the compound acts, whether the effect is mediated through microbes, host cells, or both, and whether local metabolites enter circulation. The gut–brain network explains how local chemistry can produce distributed control—and why geography may allow a multi-node molecule to be useful without allowing it to travel everywhere.

A topical wound or skin asset could offer visible endpoints and controlled exposure. It would need to distinguish antimicrobial activity from host-repair effects and show that the molecule does not simply damage cells while reducing a laboratory marker.

An anti-infective can provide a direct phenotype and an urgent need. The program would need resistance studies, pathogen spectrum, microbiome impact, manufacturing economics, and a stewardship strategy that avoids destroying the market by reserving the product only for last-line use without a sustainable model.

A metabolic asset can reach a large market and demonstrate network pharmacology. It also enters a field with powerful standards of care, long safety expectations, and intense competition. The differentiation must be measurable: durability, tolerability, tissue selectivity, lean-mass preservation, maintenance, or benefit in a defined subgroup.

A neurorepair or neurodegeneration asset may express the full ambition of state-based medicine. It will also face the longest causal chain, difficult biomarkers, species translation, and years of clinical uncertainty.

The best first asset is therefore not necessarily the largest market or the most dramatic disease. It is the program with the strongest ratio of learning to irreducible ambiguity.

A partner-ready package should answer five questions without requiring belief in the entire ghost-genome thesis:

1. **What is the defined product?** Structure, purity, stereochemistry, formulation, and resupply are controlled.
2. **What does it do?** Activity reproduces across orthogonal assays and is differentiated from nonspecific effects.
3. **Where does it go?** Relevant local and systemic exposures are measured.
4. **Why might it matter clinically?** A disease-relevant phenotype and biomarker connect the mechanism to a patient need.
5. **Why is this starting point defensible?** The source, chemical family, intellectual-property position, and manufacturing path create a practical advantage.

The platform can remain expansive behind the asset.

The asset must become specific enough to fail.

Network biomarkers

A single-target drug often carries a straightforward pharmacodynamic question: did the target move?

A state-changing therapy needs a richer but still disciplined readout. It may alter several receptors, metabolites, cell states, and microbial functions. Development cannot rely on a vague claim that the microbiome is healthier.

Network biomarkers should be specific and predeclared. They may include a defined metabolite panel, receptor-proximal signals, transcriptional signatures, barrier-permeability measures, immune-cell states, spatial imaging, or functional outputs such as glucose response, wound closure, pathogen burden, or nerve conduction.

The biomarker connects the molecular chord to the physiological state.

This is one reason the first indication matters. A platform should begin where network movement can be measured in a feasible experiment and where clinical benefit is not hidden behind years of slow progression.

Stage gates for a strange molecule

A discovery platform gains credibility when it can state in advance how a candidate earns the next dollar.

The earliest gate is physical: is the feature reproducible, distinguishable from artifact, and available in enough material for follow-up?

The next gate is structural: can identity and stereochemistry be confirmed with appropriate analytical methods and a standard?

The biological gate asks whether activity survives purification, dose-response, counterscreens, and orthogonal assays. If the effect requires a mixture, the composition and interaction must be defined.

The exposure gate asks whether the active concentration can be reached in the relevant tissue and whether metabolites preserve or destroy the effect.

The mechanism gate need not force one target, but it should identify a causal pattern precise enough to guide biomarkers, safety, and indication selection.

The developability gate addresses synthesis or fermentation, stability, formulation, selectivity, off-target liabilities, and intellectual property.

The translational gate uses a disease-relevant model with a predeclared endpoint and a comparator capable of showing whether the program is differentiated.

Only then should a candidate be treated as a serious development asset.

These gates protect both the company and the thesis. A weak molecule should be allowed to fail without becoming evidence that the hidden chemical world is irrelevant. A platform should not use its grand narrative to carry an asset that cannot pass ordinary tests.

The natural molecule as teacher

Some ghost molecules will be impossible to manufacture economically, unstable outside their niche, or unsuitable as drugs. They may still reveal a mechanism that leads to a better synthetic compound.

This is not a consolation prize. It is one of the most important routes from biology to medicine.

The natural molecule can reveal a receptor combination, an induced interface, a conformational solution, a transport strategy, or a local mechanism. Medicinal chemistry can simplify, stabilize, geofence, or amplify the relevant features.

Nature narrows the problem. Human design improves the answer.

The partnership between Ōmura and Merck illustrates this division beautifully. The microbe supplied avermectin. Industrial chemistry produced ivermectin. Pharmacology found the right use. Manufacturing created supply. Public-health partnerships created delivery.

No single participant could have completed the chain.

Intellectual property without pretending to own nature

Natural products create a recurring misunderstanding: if nature made a molecule, no one can own it.

Patent law varies by jurisdiction and fact pattern, and naturally occurring material in its natural state presents limits. Commercial protection can arise through novel derivatives, compositions, methods of use, combinations, formulations, production systems, analytical methods, and defined structural families. Trade secrets can protect source logic, fractionation, data, algorithms, and selection rules.

The strongest strategy is layered:

- **Physical access:** how material is obtained, preserved, and linked to context.
- **Discovery method:** confidential workflows connecting spectra, biology, and structural hypotheses.
- **Composition:** new molecules, analogs, and supported families.
- **Use:** disease, route, target pattern, or local mechanism.
- **Validation:** assays, biomarkers, and data packages that make the invention credible.
- **Manufacturing:** synthesis, fermentation, stabilization, and quality control.

A patent count is not a moat.

Strategic control of the chain is.

The public scientific argument must remain separate from confidential enablement. A field can understand the opportunity without being given the exact structures, algorithms, source coordinates, or unpublished assay data that make one implementation protectable.

Stewardship, access, and who benefits

A new natural-products frontier carries obligations that older discovery programs did not always meet.

Environmental microbiomes are not unlimited mines. Collection should minimize ecological impact and follow applicable laws, permits, and access agreements. Discovery material should lead to sustainable synthesis, fermentation, or heterologous production rather than repeated access from a possibly limited natural source.

Human microbiome data create privacy concerns because microbial and metabolomic patterns can reflect geography, diet, medication, disease, and identity. Population datasets need representation broad enough to avoid defining one region's chemistry as universal normality.

Benefit sharing matters as well. Biological knowledge can originate in places and communities far from the institutions that patent, finance, and commercialize the result. Legal compliance is a floor. Trust requires transparent partnerships and a serious account of who shares the risks and participates in the value created.

The Mectizan story shows one form of social translation: a commercial discovery became a long-term donation and public-private delivery system. Future models will differ, but the principle remains. A discovery fulfills its purpose only when the people who need the resulting medicine can receive it. Without durable humanitarian, public, or commercial delivery incentives, even an effective discovery may fail to reach them.

Mapping the ghost genome could improve therapeutics, ecological monitoring, agriculture, and biosecurity, but each use will require its own evidence, safeguards, and delivery architecture.

THE GHOST GENOME IN ONE VIEW

Who, where, when, and how determine what chemistry exists now.

WHO CONTRIBUTES?

- Host cells and tissues
- Bacteria and archaea
- Fungi and microbial eukaryotes
- Phages, viruses, food, drugs, and environment

WHERE DOES IT HAPPEN?

- Lumen, mucus, and biofilms
- Cell, tissue, and membrane interfaces
- Blood, organs, and circulation
- Wounds, tumors, roots, soil, and water

WHAT EXISTS NOW?

THE CONTEXT-SPECIFIC CHEMICAL EXECUTION LAYER

MAPPED REGION

Named structures, measured amounts, established pathways, and characterized functions.

GHOST REGION

Unknown, under-characterized, transient, relational, or not inferable from one genome.

IDENTITY | AMOUNT + FLUX | GEOGRAPHY | STATE | ACTIVITY

WHEN DOES IT CHANGE?

- Development and aging
- Meals, fasting, and circadian rhythms
- Stress, infection, and medication
- Injury, repair, phage lysis, and turnover

HOW IS MATTER TRANSFORMED?

- Biosynthesis and expression
- Cleavage, ligation, and cyclization
- Cross-feeding and cross-kingdom relays
- Conjugation, redox, transport, and clearance

OUTPUTS ALTER THE NEXT INPUTS, PARTICIPANTS, AND CONDITIONS.

The diagram describes the system. WHY remains a question for evidence.

THE GENOME DESCRIBES CAPACITY. CONTEXT CREATES THE CHEMICAL STATE.

Figure 19. The Ghost Genome in one view. Who contributes, where interactions occur, when the system changes, and how matter is transformed determine the context-specific chemical state. The ghost region contains unknown, under-characterized, transient, relational, or genomically irreducible chemistry. The diagram describes the system without assigning purpose.

This framework is not a claim that life can be controlled from one dashboard. It is a way to organize the questions, expose missing measurements, and create better training data for models that must learn from physical biology.

The institution is therefore part of the product. Trust influences whether samples are available, whether cohorts represent the people who will use the result, whether trials recruit, whether a therapy reaches patients, and whether benefits persist. Ethical architecture is not separate from scientific accuracy; it helps determine whether the map is correct and whether the medicine becomes real.

A global competition of architectures

The frontier-and-factory distinction does not belong to one country. But countries make choices about which frontiers become measurable.

China has become a major source of innovative pharmaceutical assets as well as integrated experimental execution. The United States retains deep strengths in universities, entrepreneurship, capital formation, pluralism, and cultural permission to challenge established categories. Europe, Japan, South Korea, Singapore, India, and other ecosystems contribute distinct capabilities.

The contest is not simply over which country has the fastest laboratory, the largest model, or the most capital.

It is over who builds the most complete learning architecture.

The Public Map Before the Private Medicine

One of the United States' recurring scientific strengths has been its willingness to build a shared map before the commercial destination was clear.

The Human Genome Project did not produce one medicine. It produced references, technologies, standards, repositories, and a common language through which thousands of later discoveries became possible. The Human Microbiome Project did something similar for the organisms living in and on the body, and its integrative phase added longitudinal multi-omic views of host and microbial activity. The Cancer Genome Atlas created multidimensional maps of tumors. The NIH Molecular Libraries Program opened large-scale small-molecule screening and deposited screening and probe data in public resources such as PubChem. LINCS cataloged cellular response signatures after small-molecule, genetic, antibody, and microenvironmental perturbations. All of Us links genomic analyses with electronic health records, surveys, physical measurements, and wearable data. HuBMAP adds spatial cellular maps. The NIH Common Fund Metabolomics program built shared resources, repositories, standards, and tools for chemical measurement.

These projects were not products.

They were observatories.

They made the previously fragmented comparable. They established common units, reference materials, data structures, and expectations. They allowed one laboratory's observation to become another laboratory's starting point. Private companies could build on a public foundation without every company first having to sequence the human genome, reconstruct the cancer atlas, or invent its own basic language for cell identity.

Together, these programs created extraordinary maps of genetic capacity, microbial membership, cellular identity, disease-associated change, perturbational response, and population variation.

Together, they still do not constitute a map of the ghost genome.

The missing object is not simply another dataset waiting to be merged. It is a physical and context-dependent chemical layer that was often never captured in the first place. A molecule may have disappeared during sample handling. It may have existed only in one layer of mucus after a meal. It may have been produced by one organism, modified by another, conjugated by the host, and consumed before reaching blood. It may be present as a reproducible activity without a confirmed structure, or as a spectrum whose family is recognizable, but whose name is absent.

Layering existing maps can reveal relationships that no single map contains.

It cannot recover physical chemistry that no program captured or preserved.

That distinction becomes urgent as the United States and other nations mobilize artificial intelligence for science. Announced in July 2026, the NIH Bio Genesis Mission aims to connect biomedical data, advanced computing, public-private partnerships, and experimental infrastructure to accelerate discovery and translation. It is a consequential architecture, but it is not yet a comprehensive ghost-genome program. Artificial intelligence can integrate only what the research system has made observable. If the chemical execution layer is missing, the model may become extraordinarily fluent in genes, proteins, diagnoses, and known metabolites while the ghost region remains outside its training experience.

A larger model cannot learn the structure-function grammar of a molecule that was never recovered.

A national ghost-genome effort would therefore have to be physical before it could become fully digital.

It would not be one central warehouse, one government laboratory, or one company surrendering its intellectual property. It would be a distributed observatory joining environmental and human sampling, biobanks, analytical centers, standards laboratories, secure population studies, synthesis capabilities, biological assays, computation, and development partners.

The public layer could establish:

- common collection and preservation metadata;
- instrument-calibration and quality-control standards;
- shared reference materials and authentic standards;
- nonproprietary spectral and molecular-family libraries;
- conventions for identification confidence;

- spatial and longitudinal sampling frameworks;
- negative-control and assay-interference datasets;
- protected prospective benchmarks for artificial intelligence; and
- rules for distinguishing a detected feature, a confirmed structure, an activity, a mechanism, and a therapeutic claim.

The proprietary frontier would retain unique samples, source histories, collection intelligence, fractionation methods, confidential algorithms, newly identified structures, analog families, assays, patents, and development data.

The public layer would make claims comparable.

The proprietary layer would give organizations a reason to take the risk of exploring.

The development factory would convert the strongest discoveries into controlled synthesis, toxicology, manufacturing, trials, and products.

This is not a static database. It is a learning loop.

A physically recovered feature receives an address. Computation proposes relationships. Biological assays reveal activity. Unexplained activity exposes a gap in the target or mechanism map. Synthesis tests the proposed structure. Negative results narrow the interpretation. The result changes what is collected, separated, measured, or tested next.

Outputs alter the next inputs, participants, and conditions.

No specialist owns the entire loop.

The loop itself is the instrument.

The global competition is therefore not over who can train the largest model on yesterday's data. It is over who can build the next layer of physical reality for every model to learn from.

A frontier without a factory produces papers and prototypes. A factory without a frontier optimizes yesterday. A public map without proprietary exploration can become an underused archive. Proprietary discovery without common standards can become a collection of incomparable claims.

The winning architecture connects all three: **a shared observatory, a protected discovery frontier, and a disciplined development factory.**

The ghost-genome opportunity will not ultimately belong to a flag. Chemistry does not recognize borders. But the nations and institutions that make this hidden layer measurable first will shape the standards, training sets, intellectual property, medicines, agricultural tools, and biological control systems built from it.

Beyond the First Medicine: Five Horizons

Therapeutics are the most immediate proof because medicine imposes a hard standard: identity, dose, safety, manufacturing, and benefit must survive in patients. But the hidden chemical language is not confined to disease.

The first horizon is therapeutics. Ghost chemistry can provide new scaffolds, target patterns, biomarkers, delivery strategies, and local interventions. The first successful product may be ordinary in form and revolutionary only in the path by which it was discovered.

The second is longevity and healthspan. Aging is not one target; it is a changing state involving repair, inflammation, mitochondrial function, immune training, barrier integrity, metabolism, senescence, and resilience. Hidden host-microbial chemistry may reveal signals associated with recovery and loss of reserve. That possibility should be explored without turning correlation into an anti-aging claim.

The third is agriculture and food security. Plant roots live inside microbial chemical neighborhoods. Signals influence germination, nutrient acquisition, drought response, pathogen resistance, growth, and soil structure. Access to currently invisible molecules could produce new crop protectants, biostimulants, diagnostics, or breeding cues while reducing dependence on broad chemical force.

The fourth is planetary health. Aquatic blooms, soils, sediments, forests, waste streams, and engineered microbiomes continuously transform carbon, nitrogen, phosphorus, metals, pollutants, and toxins. Their chemical fingerprints can become early warnings and their molecules can inspire remediation, carbon management, material recovery, or ecosystem restoration. The same caution applies: the ability to extract a signal does not grant permission to exploit the source.

The fifth is the continuation of human life in closed-loop environments. Long-duration spacecraft, lunar or Martian habitats, remote communities, and resource-limited systems cannot treat air, water, food, waste, microbes, and human health as separate utilities. They require biological systems that can be monitored, steered, and repaired before failure becomes visible at the level of symptoms.

These horizons should not be used to inflate the value of an early therapeutic program. They explain why the foundational capability - seeing chemical state in context - may matter far beyond one product.

Toward a Universal Control Layer

Engineers use a control layer to sense state, compare it with a desired range, intervene, and learn from feedback. Biology has no universal dashboard and no single master controller, but the analogy points toward a future capability.

Genomics tells us much of what a biological system can build. A context-rich chemical fingerprint can tell us more directly what the system is doing now. Repeated over time, the fingerprint can reveal trajectories: movement toward inflammation, loss of soil resilience, failure of a water community, recovery after injury, or instability in a closed habitat.

A future control layer would connect four functions: sensing the real chemical state; interpreting the state with mechanistic and statistical models; choosing the least disruptive intervention; and measuring whether the intervention moved the system toward a safer, more resilient range.

This is not a claim that one algorithm can control life, that every ecosystem has one ideal setting, or that biological complexity can be reduced to a dashboard. It is a vision of better feedback. The alternative is to keep intervening through coarse proxies while the chemistry that carries the earliest warning remains invisible.

The molecular catalog creates addresses. Activity mapping begins to show what each address can do. Unexplained activities reveal gaps in the target-and-mechanism map. Only then can a learning system begin to choose and test interventions. The Ghost Genome establishes the map required for that future: the map must come before the engine, and the engine before responsible control.

The first medicine from the ghost genome

The first medicine that validates this paradigm may not look revolutionary on the pharmacy shelf.

It may be a semisynthetic analog in an ordinary tablet. It may be a topical product, inhaled formulation, defined gut-restricted macrocycle, or diagnostic panel. The label may not mention the microbiome, the ghost genome, or four billion years of evolution.

What will make it revolutionary is the path by which it was found.

It will begin with chemistry the conventional system could not access or did not know to search. It will preserve a function that emerged from biological context. It will use artificial intelligence without mistaking prediction for evidence. It will pass the ordinary gates of identity, dose, manufacturing, toxicology, and clinical benefit.

And it will prove, in patients, that an unseen molecular signal can be translated into a controlled intervention.

One lighthouse is enough to change the coast.

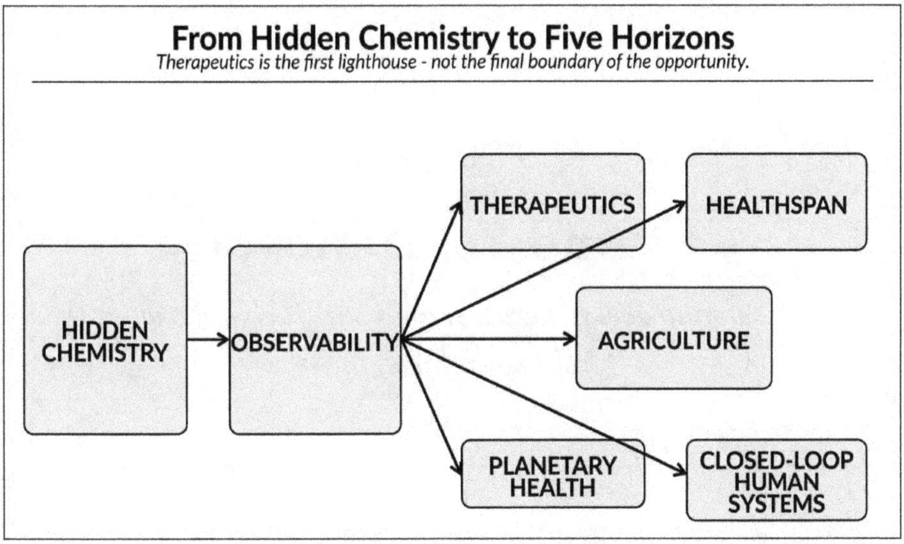

Figure 20. From hidden chemistry to five horizons: therapeutics, healthspan, agriculture, planetary resilience, and closed-loop human systems. Every application still passes through evidence gates.

EPILOGUE

And Yet It Moves

A new map does not erase the old one. It changes what the old boundaries mean.

The human genome remains foundational. Proteins remain the machinery and many of the most important targets of life. Synthetic chemistry remains one of our most powerful creative technologies. Artificial intelligence will change discovery. Peer review will remain essential. Large pharmaceutical companies will continue to do what no small laboratory can do: manufacture complex products reliably, run global trials, satisfy regulators, and deliver medicines at scale.

The argument of *The Ghost Genome* is not replacement.

It is incompleteness.

We treated durable layers of biology as though they were the whole system. We counted human genes, then microbial genes, and still expected chemistry to appear as a predictable downstream list. We built target assays and called the rest context. We filtered molecules through rules derived from the chemistry we had already succeeded in making. We used faster computers to search larger spaces without always asking whether the prevalence of viable answers was falling.

Meanwhile, the living system continued to operate.

Bile acids left the liver and returned transformed. Dietary amino acids became microbial and host-microbial metabolites. Large natural molecules folded through environments that simple property rules declared hostile. Phages accumulated at mucosal boundaries. Cells died, were consumed, and became new signals. Mass spectrometers recorded features no library could name.

And yet it moves.

The ten stories in this book began in places that looked unrelated.

A betting book at Cold Spring Harbor exposed how confidently science could misjudge the relationship between a parts count and the complexity of the whole.

A spiral bacterium in an acid stomach showed that a chronic disease could be reorganized by an observation the accepted model treated as impossible.

A star's tiny wobble revealed a giant planet where a giant planet was not supposed to be, reminding us that an instrument can force a theory to make room for reality.

A microbial culture in a household freezer preserved a molecule after a corporation had declared the program over, and the molecule later helped redraw the biology of growth and immune control.

The microbiota of twins carried aspects of metabolism into germ-free mice and revealed that a host phenotype could be transmissible, modifiable, and dependent on diet and community.

Phages in mucus became part of a protective boundary the host genome did not fully encode, showing that function can live in arrangement.

A lake turned scale from an industrial afterthought into a measurement variable.

A chip returned a microbe to its environment and showed that the laboratory can fail an organism by removing the chemistry it needs to become visible.

AlphaFold demonstrated what artificial intelligence can do when the problem has a strong representation, physical ground truth, and a benchmark that refuses self-deception.

A soil microbe near a Japanese golf course began a chain that ended in billions of treatments and the near-unthinkable possibility of eliminating diseases that had blinded and disabled people for generations.

The thread connecting the stories is not heroic rebellion. It is not that institutions are foolish, that nature is benevolent, or that outsiders possess secret access to truth.

The thread is disciplined attention to what the old map leaves unexplained.

The ghosts do not need the metaphor

If the two-billion estimate is wrong, the missing chemistry still exists at a scale we have not completed.

If cryptic assembly is rarer than I suspect, established host-microbial co-metabolism still produces molecules no single genome can fully explain.

If finite natural priors do not reverse Eroom's Law on their own, they can still improve the starting distribution for selected problems.

If a first discovery platform fails, the access problem remains.

If an instrument proves insufficient, the result should guide a better instrument or a new generation of the platform.

A good hypothesis should survive the failure of its most dramatic metaphor.

The deeper claim is simple: human biology is executed through a chemical conversation larger than our current vocabulary.

Our task is to learn the language without pretending every unknown word is profound.

That requires partitioning the unknown. Some features are artifacts. Some are known molecules in underexplored roles. Some are new structures. Some are transient states. Some will matter only to the ecology that made them. Some will become biomarkers, mechanisms, or starting points. A few may become medicines.

The standard of evidence should rise with the ambition of the claim.

A reproducible feature is not a structure. A structure is not a mechanism. A mechanism is not a therapy. A therapy is not a public-health outcome.

Ivermectin's story required every step.

Where the Map Stands Today

We are not beginning from zero. The **Microbial Metabolome Database** (MiMeDB) 2.0 currently curates 280 human–microbial co-metabolites—molecules whose natural production requires consecutive reactions from the human host and the microbiota. They include host products edited by microbes, microbial products completed by host metabolism, and dietary substrates passed through both. These molecules establish that distributed chemical authorship is not merely a theory.

But 280 is not the measured size of the ghost genome, and it is not a list of 280 uniformly reconstructed causal pathways. It is the known edge: named chemistry for which host and microbial participation has been curated. Beyond that edge are molecules with known structures but hidden activities, activities without resolved structures, incomplete routes crossing multiple organisms and kingdoms, and transient chemical states that existing catalogs were not designed to preserve.

The task is not to wait for complete understanding before beginning the catalog. The task is to give physical chemistry an address, preserve its context, interrogate its activity, map the targets and relays it exposes, and allow every result—including failure—to change the next experiment.

WHERE THE GHOST GENOME STANDS TODAY

The known edge, the unresolved frontier, and the program for moving the boundary.

THE KNOWN EDGE

280

CURATED HUMAN-MICROBIAL CO-METABOLITES

Known precedents for distributed authorship - not a complete census.

Host <-> Microbe | Diet -> Microbe -> Host | Host -> Microbe -> Host

THE MAP IS STILL INCOMPLETE

CURATED CO-METABOLITES	280
COMPLETE ROUTES	NO AUDITED COUNT
NATIVE ACTIVITY MAPS	NO AUDITED COUNT
CAUSAL TARGET + RELAY MAPS	NO AUDITED COUNT

NAME != ROUTE != ACTIVITY != MECHANISM != THERAPY

THE GHOST FRONTIER

KNOWN STRUCTURE
HIDDEN ACTIVITY

GHOST ACTIVITY
UNKNOWN MOLECULE

INCOMPLETE ROUTE
DISTRIBUTED AUTHORSHIP

TRANSIENT STATE
LOCAL OR SHORT-LIVED

MISSING RELAYS
UNKNOWN LIGANDS + NODES

CRYPTIC ASSEMBLY
MULTI-KINGDOM CHEMISTRY

THE KNOWN 280 ESTABLISH THE PRINCIPLE. THEY DO NOT DEFINE THE FRONTIER.

HOW THE BOUNDARY MOVES

1	CATALOG	Give it an address.
2	PRESERVE CONTEXT	Record who, where, when, how, and what.
3	MAP ACTIVITY	Measure what changes.
4	MAP TARGETS + RELAYS	Test how the effect propagates.
5	VALIDATE + ITERATE	Let evidence change the next experiment.

EVERY RESULT CHANGES THE NEXT COLLECTION, SEPARATION, ASSAY, OR MODEL.

Figure 21. Where the Ghost Genome stands today. MiMeDB 2.0 currently curates 280 human-microbial co-metabolites, defined as compounds requiring consecutive host and microbial biochemical reactions. They form a known edge of distributed chemical authorship, not a complete census or a uniformly route-reconstructed and causally validated set.

The next instrument is a system

No single technology will reveal the ghost genome.

Sequencing identifies capacity. Transcriptomics and proteomics reveal active machinery. Mass spectrometry detects products. Spatial methods preserve geography. Isotope tracing assigns atoms. Cultivation and community reconstruction test dependence. Synthesis confirms identity. Phenotypic assays preserve integrated function. Target assays decompose mechanism. Artificial intelligence organizes the field and selects the next experiment.

Physical scale makes rare material available.

The telescope is the connected system.

The most important feature of that system is not throughput. It is iteration. Each result should change the next sample, separation, synthesis, assay, or model. Negative results must remain in the loop. Confidence should be explicit. Hypotheses should become narrower as evidence grows, not broader.

The future of discovery will not be won by the team that produces the most predictions.

It will be won by the team that learns the most from contact with reality.

The first proof will be ordinary

Revolutions in discovery often become ordinary products.

The patient does not need a medicine to look like an ecological theory. The patient needs it to work, to be safe enough, to be manufactured consistently, and to be accessible.

The first ghost-genome medicine may be a local therapy with a conventional formulation. It may be a synthetic analog simplified from a natural scaffold. It may act through a familiar receptor in an unfamiliar combination. It may begin as a diagnostic panel that detects a chemical transition earlier than current methods. It may never carry the name of the platform that found it.

That would not diminish the discovery.

It would complete it.

The soil organism that led to ivermectin did not need to remain in the final tablet. The Rapa Nui microbe did not need to accompany sirolimus into a transplant patient. Nature provided a starting point; human systems translated it.

The same future is plausible for ghost chemistry. The natural molecule can be product, probe, or teacher.

Permission and responsibility

The work will require outsiders and insiders.

It requires entrepreneurs willing to build instruments before markets exist, and academics willing to attack those instruments until the results are trustworthy. It requires funders who can distinguish reckless speculation from a testable frontier. It requires chemists who respect evolution without romanticizing it, computational scientists who respect provenance, and clinicians who insist that system-level stories end in measurable patient benefit.

It also requires stewardship.

Environmental microbiomes are not unlimited mines. Discovery must lead to sustainable synthesis, fermentation, or other manufacturing. Source communities require ethical and legal respect. Population-scale microbiome data require privacy and fair representation. Benefits should not flow unfairly to the institutions capable of extracting value from places and communities that supplied the biological starting point.

The ghost genome is not a license to take more from nature.

It is a reason to understand what nature has already made and to reproduce useful discoveries responsibly.

What changes after the first proof

The first validated medicine or diagnostic will not complete the ghost genome. It will change the burden of proof.

Before the first lighthouse, the central argument is architectural: hidden chemistry may contain useful information that current workflows systematically miss. After a patient benefit is demonstrated, the question becomes operational: which additional ecosystems, body sites, disease states, and molecular families deserve investment?

That shift can produce its own mistakes.

Success invites imitation. Libraries will be labeled microbiome-derived without preserving meaningful context. AI platforms will claim biological grounding because one natural product appears in the training set. Unknown spectra will be promoted as proprietary assets before structures or activities are confirmed. Consumer products will borrow the language of balance and hidden signals without clinical evidence.

A field is most vulnerable to overclaiming immediately after its first genuine success.

The safeguards should therefore be built now:

- confidence levels that distinguish feature, family, structure, mechanism, and therapy;
- shared standards for sample handling and annotation;
- prospective benchmarks for AI and discovery yield;
- publication of negative and irreproducible results;
- sustainable resupply before environmental discovery becomes commercial extraction;
- clinical endpoints that matter to patients rather than only to molecular narratives;
- access plans proportionate to the public and ecological contributions behind the product.

The first proof should widen inquiry, not suspend skepticism.

The Cost of Not Looking

Not exploring the hidden chemical layer is not a neutral decision.

We already intervene in these systems. We prescribe antibiotics, immunosuppressants, metabolic medicines, diets, pesticides, fertilizers, disinfectants, and environmental controls. Each intervention changes chemistry we only partly measure. Ignorance does not preserve the natural state. It merely makes the downstream effects harder to see.

In drug discovery, the cost may be another generation of expanding search spaces and convincing false positives while biologically realized starting points remain inaccessible. In chronic disease, it may be years spent treating downstream markers while upstream coordination fails. In agriculture, it may be broad inputs that suppress a pathogen and also erase functions that supported resilience. In water and soil, it may be detecting ecosystem failure only after the visible bloom, crop loss, toxin, or collapse has arrived.

For long-duration human systems, the cost can become existential. A closed habitat cannot import a replacement biosphere after its microbial and chemical ecology drifts. The earliest useful warning may be a pattern of ghost activities long before oxygen, food production, barrier health, or waste processing crosses a familiar alarm threshold.

The danger is two-sided. Premature certainty can waste capital, harm people, and turn a young field into marketing. Passive blindness can leave consequential biology unexplored because the current instruments made absence easier to believe than invisibility.

The responsible choice is neither faith nor dismissal. It is observability: build the tools, state the confidence, test the chain, publish the failures, and let reality narrow the vision.

That is the final lesson carried by the ten stories. The discoveries became transformative because they survived the discipline that followed the surprise.

The wager was settled by annotation.

The bacterium was tested through treatment.

The planet's wobble repeated.

The freezer culture reentered formal pharmacology.

The twin phenotype changed with ecology and diet.

The mucus model was built from physical interactions.

The lake material had to become fractions, spectra, structures, and assays.

The iChip organism had to yield a real molecule.

AlphaFold faced blinded answers.

Ivermectin had to reach people, year after year.

The story creates the question.

The evidence earns the future.

The choice at the telescope

When a new instrument reveals something the old map did not predict, institutions face a choice.

They can widen the model.

Or they can explain away the signal.

The correct response is neither belief nor dismissal. It is an experiment designed to make the alternative explanations fail.

GeneSweep did not end genetics. It made genetics more interesting.

H. pylori did not make acid irrelevant. It put acid inside a fuller causal story.

51 Pegasi b did not destroy planetary science. It opened a universe of planetary architectures.

AlphaFold did not replace experimental biology. It changed which experiments became possible.

The ghost genome will not replace the genome, the proteome, medicinal chemistry, artificial intelligence, or pharmaceutical development.

It will connect them to a layer of life they have only begun to read.

The telescope is ready enough to begin.

The question is whether we will look long enough, measure carefully enough, and remain humble enough to recognize what is moving.

We have read the code.

Now we must watch the system run.

ACKNOWLEDGMENTS

To the Teachers

No book about an unfinished scientific map should pretend that one person drew it.

The central argument in these pages is assembled from work across microbiology, natural-products chemistry, pharmacology, systems biology, mass spectrometry, evolutionary biology, computational science, medicine, engineering, public health, and the study of innovation. Many of the researchers whose work made the argument possible would disagree with parts of my interpretation. That disagreement is not a flaw. It is the discipline serious inquiry requires.

I am especially grateful to the teachers whose stories anchor the book and to the many others who built the intellectual bridges among them: Matthew Bertin, Guy Carter, David Newman, William Gerrwick, Ewan Birney, Lee Rowen, Robin Warren, Barry Marshall, Michel Mayor, Didier Queloz, Carl Woese, George Fox, Suren Sehgal, Christopher Lipinski, Andrew Hopkins, Jeffrey Gordon, Vanessa Ridaura, Jeremy Nicholson, Elaine Holmes, Jeremy Barr, Forest Rohwer, Kim Lewis, Slava Epstein, Pieter Dorrestein, Michael Fischbach, John Jumper, Demis Hassabis, Satoshi Ōmura, William Campbell, Mohammed Aziz, Roy Vagelos, and the teams whose names appear in the notes and bibliography.

Their contributions gave us new instruments, new categories, and new permission to question inherited maps.

I also acknowledge the scientists, engineers, operators, collaborators, advisors, investors, and supporters who helped Biosortia pursue physical access to microbial chemistry at a scale few organizations had attempted. The work required people willing to solve problems that looked less like conventional drug discovery than environmental engineering, analytical chemistry, logistics, manufacturing, and persistence.

This manuscript deliberately omits confidential source coordinates, exact undisclosed structures, proprietary algorithms, unpublished assay details, and other information that does not belong in a public-facing book. The hypotheses are mine, as are any errors of emphasis or interpretation. The evidence belongs to the community that produced it.

APPENDIX A

A Research Agenda for the Ghost Genome

A provocative idea becomes science only when it risks being wrong.

The ghost-genome thesis contains several claims of different strength. Some are already established: microbes transform host, dietary, and pharmaceutical molecules; microbial metabolites can engage human receptors; untargeted mass spectrometry detects many features that remain unannotated; and context changes chemistry. Other claims are extrapolations: that the missing chemical layer is much larger and more functionally organized than current catalogs suggest; that distributed cross-kingdom assembly is common enough to matter broadly; and that finite natural priors can improve drug-discovery productivity.

The purpose of this appendix is to convert those extrapolations into a research program.

1. Define the counting unit before counting the ghosts

The phrase two billion ghost molecules is an order-of-magnitude hypothesis, not a measured inventory. It will remain scientifically weak unless the unit of counting is explicit.

A rigorous program should report at least four nested counts:

Exact structures. Molecules with fully resolved connectivity and stereochemistry, confirmed where possible by synthesis or an authentic standard.

Molecular entities. Distinct elemental compositions, constitutional isomers, stereoisomers, conjugates, adduct-independent metabolites, and stable covalent products that can be separated or otherwise distinguished.

Chemical states. Reversible conformers, protonation states, metal-bound states, redox states, complexes, and other forms that may alter function without constituting a new covalent structure.

Observed features. Instrument-dependent signals defined by mass-to-charge ratio, retention time, mobility, fragmentation, and spatial or temporal behavior.

These counts should never be collapsed into one number. A single molecule can produce multiple features. A feature can contain multiple coeluting molecules. A transient complex may be biologically meaningful but impossible to count as a conventional small molecule.

The hypothesis becomes testable by asking how each count scales with sampling depth, geography, diet, host state, and analytical modality.

Prediction A1: the number of reproducible chemical families will continue to rise after taxonomic and gene-catalog discovery begins to saturate

If chemistry is a combinatorial output of context rather than a simple translation of genomes, then adding time, perturbations, and spatial sampling should reveal new molecular families even when the list of organisms changes little.

A longitudinal cohort could collect stool, plasma, urine, saliva, and targeted mucosal samples under standardized dietary and medication records. The primary endpoint would be the accumulation curve for reproducible molecular families, not raw peaks. The curve should be compared with accumulation curves for species, genes, transcripts, and proteins.

A failure to observe continued chemical growth would weaken the strongest version of the thesis.

Prediction A2: many high-value unknowns will be conditionally present rather than universally rare

The usual story is that hidden molecules are missed because they exist at extremely low abundance. Some will. Others may be abundant only during short windows: after a meal, during inflammation, at a mucus boundary, following phage lysis, or during recovery.

Dense time-series sampling should identify features that are invisible in weekly or fasting-only designs. The relevant variable may be duration above a local threshold rather than mean systemic concentration.

2. Test cryptic assembly directly

The stronger cryptic-assembly hypothesis proposes that some functional molecules emerge through distributed processing involving multiple organisms, host enzymes, dietary precursors, and local chemistry. This cannot be established by correlation alone.

The decisive experiments require atom tracing and controlled reconstruction.

Prediction A3: a subset of molecules will contain atoms contributed by more than one biological kingdom or source class

Stable-isotope-labeled dietary peptides, lipids, sugars, bile acids, or amino acids can be administered in controlled systems. Host and microbial contributions can be distinguished through differential labeling, germ-free animals, defined communities, organoids, and enzyme knockouts.

A convincing example would show that:

1. a reproducible molecular feature contains labeled atoms from a dietary precursor;

2. its production requires a defined microbial organism or enzyme;
3. a host enzyme contributes a second covalent modification;
4. the complete product has a measurable phenotype or target engagement;
5. the structure is confirmed by synthesis or an orthogonal method.

One such pathway would not prove billions of distributed products. It would demonstrate that the mechanism exists.

Prediction A4: removing one community member will eliminate or redirect products not encoded as final outputs in that member's genome

Defined consortia can be assembled one member at a time. If a product depends on a relay, removing the editing organism should remove the final product while allowing an intermediate to accumulate. Adding the organism back should restore the product.

This experiment should be performed with negative controls for biomass, pH, growth rate, and substrate depletion. Otherwise, community effects can be mistaken for direct enzymatic assembly.

Prediction A5: local metals, oxygen, or redox state will change functional product distributions without changing the underlying genomes

Microfluidic mucosal models can impose controlled gradients. Metabolomic outputs can then be mapped across oxygen, pH, iron, copper, sulfur, and reactive-oxygen conditions. The aim is to identify products whose formation is partly nonenzymatic or whose binding state changes activity.

A result should be considered important only if the state is reproducible, structurally constrained, and linked to function.

3. Distinguish a molecular signal from a molecular passenger

Untargeted metabolomics is excellent at generating suspects. It is less reliable at proving guilt.

Every candidate ghost molecule should pass a causal ladder:

Association. The feature tracks with a phenotype, habitat, or state.

Identity. The molecular structure or family is established at an appropriate confidence level.

Sufficiency. Adding the molecule or a defined fraction produces a reproducible effect at plausible exposure.

Necessity. Removing the molecule, producer, enzyme, or pathway reduces the effect.

Rescue. Reintroducing the molecule or restoring the pathway restores the effect.

Exposure. The active concentration reaches the relevant compartment.

Specificity. Close analogs, inactive stereoisomers, vehicle controls, and orthogonal assays support the proposed mechanism.

Translation. The effect persists in a model that captures the disease-relevant context.

This ladder is demanding because the field is vulnerable to beautiful correlations. A feature may rise because a tissue is damaged, not because it caused the damage. A microbial metabolite may mark a diet rather than mediate its effect. A molecule may work in a cell assay at a concentration never achieved in vivo.

The research program should reward candidates for surviving the ladder, not for accumulating attractive associations.

4. Build the spatial and temporal atlas

A blood sample is a convenient national average of a highly local economy.

Many microbial and host-microbial signals act at boundaries: the mucus layer, crypt, epithelial surface, portal circulation, liver, wound edge, skin follicle, oral biofilm, or tumor margin. A systemic sample may dilute the signal or record only a downstream derivative.

Prediction A6: tissue-boundary sampling will reveal distinct molecular networks that cannot be reconstructed from stool or plasma alone

Matched spatial metabolomics, metagenomics, transcriptomics, and histology should be performed across defined tissue layers. The analysis should focus on molecular neighborhoods and gradients rather than only bulk abundance.

For example, a gut study might compare lumen, outer mucus, inner mucus, epithelium, lamina propria, portal blood, and peripheral blood. A wound study might compare necrotic center, inflammatory edge, viable tissue, exudate, and systemic circulation.

The central question is where a molecule appears, changes, and disappears.

Prediction A7: the most informative features will include trajectories, not static biomarkers

A useful signal may be a sequence: precursor rises, intermediate appears locally, host conjugate enters blood, clearance product appears in urine. Time-resolved models should represent this trajectory as one process rather than four unrelated metabolites.

Dynamic challenge studies - standardized meals, exercise, medication exposure, sleep disruption, or controlled inflammation - can reveal response capacity. The health marker may be the ability to return to a molecular baseline, not the baseline itself.

5. Test homeostatic polypharmacology as an architecture

The book argues that some natural molecules engage coordinated target sets rather than one isolated target. This must be distinguished from nonspecific promiscuity.

Prediction A8: beneficial multi-target compounds will show coherent target topology

For each candidate, the target set should be mapped across receptors, enzymes, transporters, ion channels, and signaling pathways. Network analysis should ask whether the targets converge on a physiological program, such as barrier repair, inflammatory resolution, satiety, or stress resilience.

Coherence is not proved by the number of targets. It is supported when:

- target engagement occurs within plausible free concentrations;
- target effects are complementary rather than contradictory;
- inactive analogs lose the network pattern;
- pathway perturbation changes the phenotype as predicted;
- the dose-response does not require saturating every node.

Prediction A9: some multi-node compounds will retain efficacy at lower target occupancy than matched single-target agents

This can be tested in systems where the relevant nodes and exposures are measurable. A compound that modestly engages several linked nodes may outperform a stronger single-target ligand at equal toxicity. The comparison must use matched pharmacokinetics and unbiased endpoints.

The result could also run the other way. Polypharmacology may add liabilities. The purpose is not to assume the chord is better but to measure when it is.

6. Treat geography as a drug property

A molecule's therapeutic meaning depends on where it goes.

The ghost-genome program should develop an exposure atlas for each lead: lumen, mucus, epithelium, portal blood, liver, peripheral blood, brain, kidney, and target tissue. Total concentration is insufficient; unbound concentration and active metabolites matter.

Prediction A10: gut-restricted and systemically active analogs of the same scaffold will produce separable efficacy and safety profiles

Medicinal chemistry can alter polarity, efflux, transporter recognition, metabolism, and conformational behavior. A scaffold family can therefore be used to test whether local mucosal action is sufficient or systemic exposure is required.

The study should compare:

- a gut-retained parent or analog;
- a systemically available analog;
- an inactive but exposure-matched control;
- a known pathway comparator.

This is the empirical version of geofencing.

7. Test finite natural priors in prospective portfolios

The finite-prior framework cannot be validated by citing famous natural-product drugs after the fact. It requires a prospective, matched comparison.

Prediction A11: a hybrid natural-plus-synthetic portfolio will improve validated-lead yield per dollar under defined conditions

Three channels should be established:

1. unconstrained synthetic or generative design;
2. natural-prior discovery and optimization;
3. hybrid design anchored to natural scaffolds, spectra, or mechanisms.

Programs should be matched for target class, disease difficulty, budget, and assay quality. Outcomes should include:

- number of synthesized or isolated candidates;
- confirmatory hit rate;
- false-positive rate;

- novelty;
- solubility, permeability, stability, and off-target profiles;
- time to validated lead;
- cost per validated lead;
- later-stage attrition.

The framework predicts a hybrid advantage when the natural channel meaningfully raises candidate prevalence and when access, isolation, and manufacturing costs remain manageable.

The hypothesis would be weakened if generative approaches consistently produce equal or better validated leads at lower total cost, or if natural-prior programs lose their advantage during supply and optimization.

8. Establish the minimum standard for a ghost-molecule claim

The field needs language that prevents speculation from hardening into fact.

A publication or company presentation should report:

- the identification confidence level;
- whether an authentic standard exists;
- exact sample handling and extraction;
- prevalence and reproducibility across batches or people;
- evidence for microbial, host, dietary, or environmental origin;
- measured exposure in the relevant compartment;
- causal evidence;
- alternative explanations;
- negative data;
- whether the molecule is a discovery marker, mechanistic probe, or therapeutic candidate.

The phrase ghost molecule should not become a permission slip for vague claims. It should identify an unknown with a plan for becoming known.

9. Build shared infrastructure without erasing proprietary incentive

A field this complex needs reference standards, shared spectra, common metadata, and reproducible protocols. It also requires investment in expensive access, isolation, synthesis, and validation.

The balance can be achieved through layers.

Public infrastructure should include annotation standards, negative-control datasets, common reference materials, instrument calibration practices, and nonproprietary spectral libraries. Proprietary programs can protect unique samples, source histories, isolation methods, structures, analogs, assays, and development data.

The public layer makes claims comparable. The proprietary layer gives organizations a reason to build the frontier.

10. State in advance what would change the thesis

The ghost-genome argument should be revised if future evidence shows that:

- most persistent unknown features are technical artifacts, adducts, or derivatives of a relatively small known set;
- chemical-family discovery rapidly saturates once sampling and annotation improve;
- distributed cross-kingdom assembly is rare and functionally marginal;
- unknown metabolites contribute little causal information beyond known pathways;
- natural priors do not improve validated-lead yield after total cost is counted;
- network pharmacology adds more toxicity than efficacy in relevant comparisons;
- proposed diagnostic patterns fail replication across cohorts and laboratories.

A theory that cannot name its failure conditions is branding, not science.

The research agenda is ambitious, but it is divisible. No single experiment must prove a two-billion-molecule orchestra. The work can proceed one atom, one pathway, one tissue boundary, and one reproducible causal chain at a time.

That is how ghosts become chemistry.

APPENDIX B

How to Think About the Two-Billion Hypothesis

The number two billion is deliberately provocative. It is also easy to misunderstand.

It is not a claim that two billion stable, fully resolved, simultaneously present molecules have been counted in one person. It is a hypothesis about the potential output space of human-associated chemistry across people, microbial communities, diets, tissues, developmental stages, disease states, medications, and time.

The estimate emerged from a simple concern: conventional molecule counts often assume a nearly linear relationship between genes, pathways, and products. Living chemistry is not linear.

This appendix offers a transparent thought experiment rather than a proof.

Start with molecular families, not genes

The human microbiome contains enormous genetic diversity at the population level. Specialized-metabolite pathways, primary metabolism, transport, stress responses, proteolysis, and host enzymes can create or modify chemical substrates. Yet gene count alone is a poor estimator because many genes do not produce small molecules, while one pathway can produce many analogs.

A more useful starting point is the number of reproducible molecular families that can be physically detected across human-associated and relevant environmental microbiomes.

Suppose future integrated surveys identify one million distinct families across the global population, diet, environment, and disease states. That number is not established; it is a placeholder for the thought experiment.

If each family contains, on average, twenty stable covalent variants across substrate choices, tailoring, stereochemistry, oxidation, conjugation, and host transformation, the count becomes twenty million entities.

If those entities appear in an average of ten biologically distinct bound, redox, metal-associated, complexed, or compartment-specific states, the functional state space becomes two hundred million.

If time, diet, medication, phage activity, and community composition generate an average of ten reproducible context-specific products or transient intermediates per state across the human population, the order of magnitude reaches two billion.

The arithmetic is not the evidence. It demonstrates that the number is not numerically absurd once combinatorial context is admitted.

Why the estimate could be too high

Several factors can inflate the count:

• mass-spectrometry adducts and fragments can be mistaken for distinct molecules;

• the same molecule can appear under multiple retention times or ion forms;

• transient states may not deserve to be counted as separate entities;

• many theoretical combinations will never occur;

• chemical families may contain fewer functional variants than assumed;

• annotation improvements may collapse thousands of features into known pathways;

• rare, person-specific molecules may have no meaningful biological role.

A serious estimate must remove these redundancies.

Why the estimate could be too low

Other factors can hide diversity:

• most studies sample stool or blood rather than tissue boundaries;

• fasting samples miss post-meal and circadian chemistry;

• extraction methods cover only part of chemical space;

• unstable, reactive, nonionizing, metal-bound, and membrane-associated molecules are underdetected;

• global diets and medications create vast substrate diversity;

• phages, fungi, archaea, and small eukaryotes are less characterized than bacteria;

• the same nominal molecule can have different biological meaning by stereochemistry, location, complexation, and timing;

• the global population contains far more microbiome diversity than any current cohort.

The purpose of the estimate is therefore not precision. It is to challenge a hidden assumption that the unknown chemical universe must be small because current catalogs are small.

A better future number

A scientifically mature estimate would be reported as a range with explicit categories:

• confirmed unique structures;

• high-confidence structures;

• molecular families;

- reproducible unknown features;
- inferred transient or complexed states;
- predicted but unobserved entities.

It would include accumulation curves and uncertainty intervals. It would distinguish one-person diversity from population diversity and simultaneous abundance from lifetime output.

The two-billion hypothesis has done its job if it motivates that accounting.

The number should not become sacred. The missing chemistry should become measurable.

REFERENCE

Glossary of Terms and Abbreviations

This glossary is written for readers crossing several fields at once. Definitions are intentionally fuller than dictionary entries and, where relevant, explain how the term is used in this book.

16S rRNA sequencing

A method that identifies and compares bacteria and archaea by sequencing regions of the gene encoding the small-subunit ribosomal RNA. Because this gene contains both conserved and variable regions, it became a foundational tool for microbial classification and microbiome surveys. It usually provides less functional detail than whole-metagenome sequencing.

ABC transporter

An ATP-binding cassette transporter. These membrane proteins use energy from ATP to move molecules across cell membranes. In pharmacology, ABC transporters can limit drug absorption, remove compounds from cells, protect tissues, or contribute to multidrug resistance.

Absorption, distribution, metabolism, and excretion (ADME)

The major processes that determine what the body does to a drug or other molecule. Absorption describes entry into the body; distribution describes where it goes; metabolism describes chemical transformation; and excretion describes removal. Toxicity is often added to form ADMET.

Activity-based protein profiling (ABPP)

A chemical-proteomic approach that uses reactive or affinity probes to measure the functional state of enzymes and other proteins in complex biological samples. Unlike a simple abundance measurement, ABPP asks which proteins are active or engaged.

AlphaFold 3

A deep-learning system reported in 2024 for predicting joint structures of biomolecular complexes that can include proteins, nucleic acids, small molecules, ions, and modified residues. Its success illustrates how AI benefits from physical structure, curated examples, and a clearly defined prediction target.

Active learning

A machine-learning strategy in which a model chooses the next experiment expected to reduce uncertainty most efficiently. In drug discovery, active learning can prioritize which molecule to synthesize, assay, or characterize next.

Adduct

A noncovalent or covalent association between a molecule and another species. In mass spectrometry, sodium, potassium, ammonium, or solvent adducts can make one molecule appear as several signals. Correctly distinguishing adducts from unique molecules is essential to estimating chemical diversity.

Aryl hydrocarbon receptor (AhR)

A ligand-activated transcription factor that responds to many environmental, dietary, host, and microbial compounds, including some tryptophan-derived metabolites. AhR helps regulate barrier tissues, immune responses, and detoxification pathways. Its effects depend on ligand, tissue, dose, and timing.

Artificial intelligence (AI)

A broad term for computational systems that perform tasks associated with prediction, pattern recognition, generation, reasoning, or decision support. In this book, the important distinction is between AI that generates possibilities and AI that is constrained by physical biological evidence.

Autophagy

A conserved cellular recycling process in which components are enclosed and delivered to lysosomes for degradation. Autophagy removes damaged organelles and proteins, releases reusable building blocks, and influences stress responses, immunity, and metabolism.

Bacteriocin

A peptide or protein produced by bacteria that inhibits other bacteria, often within a relatively narrow taxonomic range. Microcins are a well-known class of small bacteriocins. Bacteriocins can shape microbial communities and may have therapeutic potential.

Bacteriophage, or phage

A virus that infects bacteria. Phages can kill bacterial cells, integrate into bacterial genomes, move genes, alter community structure, and become enriched at mucosal surfaces. The collection of phages associated with humans is part of the human virome.

Bacteriophage adherence to mucus (BAM)

A model proposing that some phages bind to mucus and provide a non-host-derived layer of defense by increasing encounters with bacterial hosts at mucosal surfaces. The concept illustrates how biological function can emerge from spatial arrangement rather than one genome alone.

Base rate

The underlying prevalence of a true condition or viable candidate in the population being searched. Even a highly accurate test can generate mostly false positives when the base rate is extremely low. The finite-prior argument in this book is partly an argument about improving the base rate of useful chemical starting points.

Beyond the Rule of Five (bRo5)

Chemical space outside the conventional property ranges associated with Lipinski's Rule of Five, often including larger, more flexible, or more polar molecules such as macrocycles. Some bRo5 compounds achieve oral or cellular exposure through conformational adaptation, intramolecular hydrogen bonds, active transport, or other mechanisms.

Breakthrough aggressiveness

The disciplined willingness to challenge the framing of a problem, build a missing instrument, and pursue high-consequence uncertainty while applying conservative standards to claims and evidence. It does not mean moving recklessly or lowering the burden of proof.

Applicability domain and coverage bias

The region of chemical and property space for which a model has relevant training support. Coverage bias occurs when datasets sample that space unevenly, allowing apparently strong performance on familiar examples while predictions remain unreliable for underrepresented structures or biological states.

Bioactivity

A measurable effect of a molecule or mixture on a biological system. Bioactivity may include binding to a target, changing an enzyme, altering gene expression, affecting cell behavior, inhibiting an organism, or improving a disease-relevant phenotype. Bioactivity is not the same as clinical benefit.

Bioavailability

The fraction of an administered dose that reaches systemic circulation in an active form. Oral bioavailability reflects absorption and first-pass metabolism. A gut-restricted therapy may be intentionally effective with little or no systemic bioavailability.

Biofilm

A structured community of microorganisms embedded in a self-produced matrix and attached to a surface or interface. Biofilms change microbial physiology, protect cells from stress, alter chemical gradients, and can be beneficial, neutral, or pathogenic depending on context.

Biological ground truth

A phrase used in this book for traceable physical evidence linking material, measurement, context, and function. It can include a recovered molecule, its spectrum, source conditions, biological activity, and confirmation. The phrase contrasts with a prediction that has not yet been connected to physical biology.

Bioreactor

An engineered vessel or system that supports biological processes under controlled conditions. The book uses human bioreactor as a metaphor for the continuously operating host-microbial ecosystem that transforms food, host products, microbial products, medicines, and cellular debris. The body is not literally governed like an industrial reactor, and the metaphor should not erase agency, anatomy, or evolution.

Biosynthetic gene cluster (BGC)

A group of neighboring genes that collectively encode enzymes, regulators, transporters, or resistance functions associated with production of a specialized metabolite or metabolite family. BGCs are valuable predictors of chemical capacity but do not always specify the final molecule produced in a living community.

Central dogma of molecular biology

Francis Crick's framework describing information transfer from nucleic acid to nucleic acid and from nucleic acid to protein, while rejecting information transfer from protein back to nucleic acid or protein. It is often simplified as DNA to RNA to protein. The book criticizes an overly narrow cultural interpretation of the dogma, not the molecular principle itself.

Chameleonicity

The ability of a molecule to change conformation and expose or shield polar surface area depending on its environment. Chameleonic behavior can help some large molecules remain soluble in water yet permeate membranes.

Chemical ecology

The study of how chemicals mediate interactions among organisms and their environments. Chemical ecology emphasizes concentration, location, timing, producer, receiver, and ecological consequence rather than treating a molecule as having one fixed purpose.

Chemical execution layer

The changing molecular state produced when genetic capacity, cellular machinery, microbes, viruses, food, medicines, environment, anatomy, and time interact. The term emphasizes what a living system is doing now rather than only what its genomes could do.

Chemical feature

An instrument-defined signal, often characterized by mass-to-charge ratio, retention time, ion mobility, isotope pattern, fragmentation, and abundance. A feature is not automatically a unique molecule. One molecule can generate several features, and one feature can contain several compounds.

Chemical family

A group of structurally related molecules sharing a scaffold, substructure, biosynthetic origin, or fragmentation pattern. Molecular networking can reveal families even when most members remain unidentified.

Chemoproteomics

The use of chemical probes and proteomic methods to identify protein targets, binding partners, or functional states across many proteins at once. Chemoproteomics can help connect molecules to mechanisms in cells or tissues.

Clinical attrition

The loss of drug candidates during clinical development because of insufficient efficacy, unacceptable safety, poor exposure, manufacturing problems, trial design, commercial decisions, or other causes. Late-stage attrition is especially expensive because substantial capital and time have already been spent.

Commensal

An organism living in or on a host without necessarily causing harm. The traditional definition implies one organism benefits while the other is unaffected, but microbiome relationships often shift among commensalism, mutualism, and harm depending on context.

Compound-first discovery

A discovery strategy that begins with a physically captured chemical feature or purified compound rather than first identifying and culturing the producing organism. Compound-first approaches can reveal chemistry from complex environments but may make producer assignment and resupply more difficult.

Conformer

One of several three-dimensional shapes accessible to a molecule through rotation or other movement without changing covalent connectivity. Different conformers can differ in permeability, target binding, and metabolism.

Conjugation

The covalent attachment of a chemical group to a molecule. Host enzymes commonly add glucuronide, sulfate, acetyl, methyl, glycine, taurine, or other groups. Microbes also add and remove conjugates. Conjugation can change solubility, transport, receptor activity, and clearance.

Connective tissue and extracellular matrix (ECM)

The structural network of proteins, polysaccharides, and associated molecules surrounding cells. The ECM is not inert scaffolding; it stores signals, controls mechanics, and changes during injury, repair, fibrosis, and disease.

Cryptic assembly

The author's hypothesis that some functional molecules emerge through distributed processing involving microbial enzymes, host enzymes, dietary or cellular precursors, local metals, redox conditions, and transient chemistry. The final product may not be explicitly encoded by any one genome or predicted from a single BGC.

Cytochrome P450 (CYP450)

A large family of heme-containing enzymes that oxidize drugs, lipids, steroids, and other molecules. CYP enzymes are prominent in the liver and intestine but also occur elsewhere. They can activate, deactivate, or toxify compounds.

Dalton (Da)

A unit of molecular mass is approximately equal to the mass of one proton or neutron. Molecular weights in this book are often given in daltons. A 750-Da molecule is larger than many conventional oral drugs but smaller than most proteins. The book uses approximately 750 Da as an illustrative exploration frontier, not as a universal definition of a small molecule or an upper limit on biological signaling. Real signaling molecules occur throughout and beyond this range.

Dark matter of metabolomics

A metaphor for the many reproducible mass-spectrometry signals that cannot be confidently assigned to known structures. Some unknowns are biologically important; others are artifacts, adducts, fragments, contaminants, or variants of known chemistry.

Dysbiosis

A change in microbial community structure or function associated with disease or impaired resilience. Dysbiosis should not be treated as one universal condition. The same taxonomic change can have different meanings across hosts, diets, treatments, and environments.

Endogenous

Produced within an organism or system. In host-microbiome biology, authorship can be distributed, so a molecule present within the body may be host-derived, microbially derived, diet-derived, co-metabolized, or environmentally introduced.

Enteric nervous system (ENS)

The network of neurons and supporting cells embedded in the gastrointestinal tract. It controls motility, secretion, blood flow, and local reflexes and communicates bidirectionally with the brain, immune system, endocrine system, and microbes.

Enterohepatic circulation

The cycling of compounds between the liver, bile, intestine, and back to the liver through portal blood. Bile acids are a central example. Microbial transformations during this cycle create compounds with different signaling properties.

Eroom's Law

The observation that inflation-adjusted pharmaceutical research and development productivity declined for decades, roughly halving in terms of new drugs per billion dollars every nine years in the historical analysis by Scannell and colleagues. The name is Moore's Law spelled backward.

Evolutionary prior

A starting assumption informed by the fact that a molecule, scaffold, pathway, or interaction has been physically realized in living systems. An evolutionary prior is not proof of safety or therapeutic value. It narrows the search by beginning with chemistry that biology has made.

False-positive rate

The proportion of negative cases incorrectly classified as positive by a test or screen. In low-prevalence searches, even a modest false-positive rate can overwhelm the true positives.

Finite natural prior

The author's term for a physically realized molecule, scaffold, spectrum, or biological pattern recovered from a finite natural system and used to guide discovery. The proposed value is enrichment, not infallibility.

Food-derived bioactive peptide

A peptide released from a dietary protein by digestion, fermentation, processing, or microbial enzymes that has biological activity beyond serving as a simple amino-acid source. Evidence is strong for some peptides in vitro and more limited for many in vivo claims because stability and exposure are difficult.

Gnotobiotic

Describing an animal whose microbial status is known. Germ-free animals contain no detectable microorganisms under defined conditions; other gnotobiotic animals carry a specified community. These models are powerful for causal studies but do not fully reproduce normal human ecology.

G-protein-coupled receptor (GPCR)

A large family of cell-surface receptors that convert extracellular signals into intracellular responses through G proteins and related pathways. GPCRs respond to hormones, neurotransmitters, lipids, odors, metabolites, and some microbial products and are major drug targets.

Geofencing

The intentional restriction of a molecule's exposure to a tissue or compartment, such as the gut lumen or mucosa. A geofenced drug aims to achieve local efficacy while reducing systemic or central nervous system exposure and associated toxicity.

Germ-free

Raised and maintained without detectable microorganisms. Germ-free models reveal microbial contributions by comparison with colonized models, but their immune systems, metabolism, and anatomy develop differently from conventionally colonized animals.

Global Natural Products Social Molecular Networking (GNPS)

An open computational ecosystem for sharing, comparing, networking, and annotating tandem mass-spectrometry data. GNPS helps organize unknown spectra into related families and supports community curation.

Ghost activity

A reproducible biological effect whose responsible chemistry is not yet resolved. A ghost activity can guide fractionation and experiment design, but it should not be treated as a confirmed molecule, target, or mechanism.

Ghost genome

The author's working name for the dynamic, partly unobserved chemical execution layer produced by host, microbial, viral, dietary, environmental, and temporal interactions. It is not a literal chromosome and does not replace genomics.

Great Plate Count Anomaly

The long-recognized gap between the number of microbial cells observed in natural samples and the much smaller number that form colonies under standard laboratory culture conditions. The anomaly helped motivate culture-independent and in situ cultivation methods.

Gut-brain axis

The bidirectional communication network linking the gastrointestinal tract and central nervous system through neural, endocrine, immune, vascular, and metabolic routes. Microbial effects on the brain are plausible and sometimes demonstrated, but the phrase is often overused in consumer claims.

Holobiont

A host and its associated microorganisms considered as an ecological unit. The term can be useful for emphasizing interaction but should not imply that host and microbes always share one evolutionary interest or behave as a single organism.

Homeostasis

The dynamic maintenance of physiological variables within viable ranges. Homeostasis is not a fixed equilibrium; it involves sensing, compensation, feedback, and adaptation.

Homeostatic polypharmacology

A term used in this book for multi-target activity that coherently supports a physiological program rather than causing unrelated off-target effects. It is a hypothesis to be demonstrated through target topology, exposure, dose-response, and causal experiments.

Horizontal gene transfer (HGT)

Movement of genetic material between organisms outside direct parent-to-offspring inheritance. Transformation, conjugation, transduction, mobile elements, and viruses can move genes across microbial lineages, including resistance and metabolic functions.

Human Microbiome Project (HMP)

A U.S. National Institutes of Health program launched in 2007 to characterize microbial communities associated with healthy humans and develop methods and reference datasets. It helped establish the microbiome as a major field of biomedical research.

Human virome

The collection of viruses associated with the human body, including viruses that infect human cells and the far more numerous bacteriophages that infect bacteria. Viromes vary substantially across individuals and body sites.

iChip

An in situ cultivation device containing many small chambers that allow individual microbial cells to grow while exposed to chemical conditions from their natural environment. The approach helped recover previously uncultured organisms and contributed to discovery of teixobactin.

Indole

An aromatic ring system present in tryptophan and many biologically active molecules. Gut microbes convert tryptophan into indole and numerous derivatives that can affect microbial behavior and host receptors.

Indole-3-propionic acid (IPA)

A microbial metabolite produced from tryptophan by certain gut bacteria. IPA circulates in humans and has been studied for antioxidant, barrier, metabolic, and receptor-related effects. Its biology is active and context-dependent rather than reducible to one mechanism.

Industrial-Scale Microbiome Mining (ISMM)

Biosortia's term for recovering and fractionating microbial community material at unusually large physical scale to access low-abundance chemistry produced in situ. In this book, ISMM is presented as one proposed discovery instrument and a source of biological ground truth, not as proof of every ghost-genome claim.

Ion mobility

A gas-phase separation technique that distinguishes ions by their movement through a buffer gas under an electric field. Ion mobility adds information about size, shape, and charge and can help separate isomers or conformers in mass spectrometry.

Isomer

A molecule with the same molecular formula as another molecule but a different arrangement of atoms or a different three-dimensional configuration. Isomers can have very different biological activities and toxicities.

Last Universal Common Ancestor (LUCA)

The inferred ancestral population from which all known cellular life descends. LUCA was not necessarily the first life or a single individual. The concept highlights deep conservation in translation, metabolism, and cellular chemistry.

LC-MS/MS

Liquid chromatography-tandem mass spectrometry. Liquid chromatography separates molecules; the mass spectrometer measures ions; and tandem mass spectrometry fragments selected ions to obtain structural information. LC-MS/MS is central to metabolomics, natural-products discovery, and bioanalysis.

Ligand

A molecule that binds to another molecule, often a receptor, enzyme, protein, metal, or nucleic acid. Binding alone does not establish that the ligand activates, inhibits, or meaningfully changes the target in vivo.

Lipinski's Rule of Five

A set of empirical guidelines associated with oral absorption and permeation for many conventional small molecules: molecular weight no greater than about 500, lipophilicity within a specified range, and limited hydrogen-bond donors and acceptors. The rules were intended as practical guidance, not universal laws.

Lipopeptide

A molecule containing a peptide linked to a lipid. Lipopeptides can interact with membranes, immune receptors, transport systems, and microbial cells. They range from signaling molecules to antibiotics and toxins.

Macrocycle

A molecule containing a large ring, commonly defined in medicinal chemistry as twelve or more atoms in the ring. Macrocycles can preorganize binding groups, resist degradation, engage broad protein surfaces, and sometimes display chameleonic permeability.

Markush structure

A patent convention that represents a family of related chemical structures using variable substituents and alternatives. A well-supported Markush claim can protect a chemical neighborhood rather than one exact molecule, but it must satisfy legal requirements for written description and enablement.

MassQL

Mass Spec Query Language, a platform-independent language for expressing and searching mass-spectrometry patterns such as diagnostic fragments, neutral losses, isotope signatures, and chromatographic constraints. It helps researchers retrieve related spectra without requiring a complete library match.

Mass defect

The small difference between an isotope's exact mass and its nominal whole-number mass. Accurate mass defects help infer elemental composition and distinguish molecular classes in high-resolution mass spectrometry.

Mass spectrometry (MS)

An analytical technique that converts molecules into ions and measures their mass-to-charge ratios. Different ionization, separation, and fragmentation methods reveal different parts of chemical space.

Metabolite

A molecule produced, consumed, or transformed in metabolism. The term includes central metabolites, signaling molecules, xenobiotic products, microbial compounds, host-microbial co-metabolites, and degradation products.

Metabolome

The collection of metabolites in a defined biological system, sample, compartment, or time. Unlike a genome, a metabolome changes rapidly with diet, drugs, environment, development, and disease.

Metabolomics

The systematic measurement and analysis of metabolites. Targeted metabolomics quantifies predefined compounds, while untargeted metabolomics surveys a broader set of features and generates discovery hypotheses.

Metagenomics

The sequencing and analysis of DNA recovered directly from a community rather than from one isolated organism. Shotgun metagenomics can reveal genes, pathways, and genomes but does not by itself show which genes are active or which molecules are present.

Metaproteomics

The study of proteins produced by a microbial community. Metaproteomics brings genetic capacity closer to function but remains technically challenging because of shared peptides, dynamic range, and incomplete reference databases.

Metatranscriptomics

The study of RNA transcripts produced by a microbial community at a particular time. It indicates active gene expression more directly than DNA abundance but does not guarantee protein or metabolite production.

Microbiome

The microorganisms in a defined habitat together with their genes, products, and ecological context. In popular use, microbiome is often used interchangeably with microbiota, which more narrowly refers to the organisms themselves.

Microcins

Small antibacterial peptides produced by some Enterobacteriaceae. They can inhibit closely related bacteria and influence competition in the gut. Their narrow targeting has inspired therapeutic interest.

Molecular networking

A computational method that connects tandem mass spectra based on similarity, helping organize related molecules into families. A network edge suggests related fragmentation, not definitive structural identity.

Multi-omics

The combined analysis of two or more molecular layers, such as genomics, transcriptomics, proteomics, metabolomics, epigenomics, or spatial data. Integration is powerful but can multiply noise if each layer lacks quality and causal design.

Multi-node control

Regulation achieved by influencing several connected components of a biological network. Multi-node control can be produced by one polypharmacological molecule, a mixture, or a sequence of signals.

Natural product

A chemical substance produced by a living organism. In drug discovery, natural products include microbial, plant, marine, and animal molecules and their derivatives. The category spans benign nutrients, signaling compounds, toxins, and medicines.

N-acyl amide

A molecule containing an acyl group linked to an amine. Some bacterial N-acyl amides resemble mammalian signaling lipids and can interact with human GPCRs.

Nonenzymatic chemistry

Chemical reactions that occur without direct enzyme catalysis, including oxidation, hydrolysis, condensation, rearrangement, metal-mediated reactions, and spontaneous adduct formation. In living systems, nonenzymatic reactions are shaped by local pH, redox state, metals, concentration, and confinement.

Nonribosomal peptide synthetase (NRPS)

A large modular enzyme system that builds peptides without the ribosome. NRPS pathways can incorporate noncanonical amino acids, modify residues, and generate cyclic or branched structures.

Nuclear magnetic resonance spectroscopy (NMR)

An analytical method that uses the magnetic behavior of atomic nuclei to determine molecular structure, stereochemistry, interactions, and dynamics. NMR usually requires more material than mass spectrometry but provides powerful structural confirmation.

Nuclear receptor

A receptor that regulates gene transcription in response to ligands such as steroids, lipids, bile acids, vitamins, and xenobiotics. Some nuclear receptors also respond to microbial or host-microbial metabolites.

Off-target effect

An effect caused by interaction with a molecule other than the intended target. Off-target effects can be harmful, neutral, or occasionally beneficial. They should be distinguished from intended multi-target pharmacology.

Observability

The degree to which a physical or biological phenomenon can be detected, preserved, distinguished from noise, identified, and tested under conditions relevant to its function. Observability depends on the entire measurement chain, not sensitivity alone.

Orphan biosynthetic gene cluster

A BGC whose associated small-molecule product is unknown. Orphan clusters reveal biosynthetic potential that has not yet been connected to physical chemistry.

Pangenome

The complete collection of genes found across all sampled strains of a microbial species. The pangenome includes a **core genome** shared by nearly all strains and an **accessory genome** containing genes present in only some strains. Accessory genes can encode nutrient use, drug metabolism, virulence, resistance, transport, ecological competition, phage defense, and specialized-molecule production. A pangenome describes the genetic capabilities distributed across a species; it does not establish which capabilities are active or which molecules are present in a particular biological state.

Peptidase or protease

An enzyme that cleaves peptide bonds. Proteases digest food, process signaling proteins, remodel tissue, activate or deactivate peptides, and contribute to microbial competition. The terms are often used interchangeably, though peptidase can emphasize smaller peptide substrates.

Phenotypic screening

Testing compounds for an observable effect in cells, tissues, organisms, or disease models without requiring a predefined single target. Phenotypic screening can preserve network context but may make mechanism identification more difficult.

Phage lysis

The rupture of a bacterial cell caused by phage replication or phage-encoded enzymes. Lysis releases new phage particles and bacterial contents, creating a local chemical event.

Pharmacodynamics (PD)

What a drug does to the body or biological system, including target engagement, pathway effects, biomarkers, efficacy, and toxicity as a function of concentration and time.

Pharmacokinetics (PK)

What the body does to a drug, usually described by concentration over time, absorption, distribution, metabolism, and excretion. PK determines whether a molecule reaches a target at an effective and safe exposure.

Phenotype

An observable trait or state produced by the interaction of genes, environment, development, and history. In cells, a phenotype may be morphology, viability, signaling, differentiation, or another measurable behavior.

Polyketide synthase (PKS)

A modular enzyme system that constructs polyketides from small carbon building blocks. PKS pathways produce diverse natural products, including many antibiotics, immunosuppressants, and toxins.

Polypharmacology

The activity of one molecule at multiple biological targets. Polypharmacology can be intentional and therapeutically coherent, or it can produce unwanted side effects. The quality of the target pattern matters more than the target count.

Positive predictive value (PPV)

The probability that a positive test or screen result is a true positive. PPV depends on test performance and the prevalence of true positives in the population being searched.

Postbiotic

A preparation of nonliving microorganisms and/or their components that confers a health benefit on the host under the consensus definition developed by the International Scientific Association for Probiotics and Prebiotics. In commercial use, the word is sometimes applied more broadly to microbial metabolites; the exact meaning should be stated.

Prebiotic

A substrate selectively used by host microorganisms that confers a health benefit. Not every dietary fiber is a prebiotic, and the effect depends on the recipient community and dose.

Probiotic

A live microorganism that, when administered in adequate amounts, confers a health benefit on the host. Species name alone is insufficient; benefits are strain- and context-specific.

Proteome

The full set of proteins expressed in a cell, organism, community, or sample under defined conditions. Unlike the genome, the proteome changes with time and state.

Pregnane X receptor (PXR)

A nuclear receptor that senses many endogenous molecules and xenobiotics and regulates transport and detoxification genes. Some microbial metabolites influence PXR-associated barrier and metabolic pathways.

Quantitative estimate of drug-likeness (QED)

A composite score that summarizes how closely a molecule resembles property distributions observed in approved drugs. QED can aid prioritization but can penalize valid compounds outside conventional chemical space.

Reductionism

The strategy of understanding a system by decomposing it into parts. Reductionism is indispensable for causal science. The problem arises when findings from isolated parts are assumed to fully explain an interacting system.

RORDEP

Ruminococcus torques-derived peptide. RORDEP1 and RORDEP2 are two bacterial polypeptides reported in 2025, detected in human plasma and studied for metabolic effects in rodents. Their human physiological and therapeutic significance remains unproven.

Resilience

The capacity of a system to absorb disturbance, maintain function, or return to a viable state. In microbiome research, resilience can be more informative than one static definition of a normal community.

Ribosomally synthesized and post-translationally modified peptide (RiPP)

A peptide made initially by the ribosome and then extensively modified by enzymes. RiPPs can contain rings, unusual linkages, dehydrated residues, lipids, or other features and include many microbial signaling and defense molecules.

Rule of Five

See Lipinski's Rule of Five.

Secondary bile acid

A bile acid generated when gut microbes modify host-produced primary bile acids through deconjugation, dehydroxylation, epimerization, and other reactions. Secondary bile acids can affect microbes, membranes, and host receptors.

Secondary metabolite or specialized metabolite

A molecule not required for universal core metabolism but important for ecological functions such as signaling, competition, defense, metal acquisition, development, or symbiosis. The term specialized metabolite avoids implying that these compounds are biologically secondary in importance.

Semisynthesis

Chemical modification of a molecule originally produced by a biological system. Semisynthesis can improve potency, exposure, stability, safety, manufacturing, or intellectual-property position while preserving a valuable natural scaffold.

Sensitivity

The proportion of true positives correctly detected by a test or screen. Sensitivity must be interpreted with specificity, false-positive rate, and prevalence.

Short-chain fatty acid (SCFA)

A small fatty acid, commonly acetate, propionate, or butyrate, produced in substantial amounts by microbial fermentation of dietary substrates. SCFAs serve as fuels and signals and can affect epithelial, immune, metabolic, and neural pathways.

Small Molecule In Situ Resin Capture (SMIRC)

A culture-independent method that places adsorbent resin in a natural environment to capture released organic compounds where they are produced. SMIRC is a compound-first access strategy and samples a different chemical fraction from cultivation or bulk biomass recovery.

Silent biosynthetic gene cluster

A BGC that is present in a genome but minimally expressed or inactive under the tested conditions. Changes in nutrients, stress, competitors, host signals, or regulation can activate silent pathways.

Solute carrier (SLC)

A large family of membrane transport proteins that move nutrients, metabolites, ions, and drugs. SLC transporters can enable tissue uptake or influence drug disposition.

Spatial metabolomics

The mapping of metabolites within intact or minimally disrupted tissue and surfaces. Imaging mass spectrometry and related methods reveal where molecules are produced, retained, consumed, or transformed.

Specialized pro-resolving mediator (SPM)

A family of lipid mediators, including resolvins, protectins, and maresins, implicated in active resolution of inflammation. Their biology illustrates that ending inflammation is an organized process, although measurement and translation remain active areas of research.

Stable-isotope tracing

The use of nonradioactive isotopes, such as carbon-13 or nitrogen-15, to follow atoms through metabolic pathways. Mass shifts reveal which products contain atoms from a labeled precursor.

Stereochemistry

The three-dimensional arrangement of atoms in a molecule. Molecules with the same connectivity but different stereochemistry can have different targets, potency, metabolism, and toxicity.

Structure elucidation

The process of determining a molecule's elemental composition, connectivity, functional groups, stereochemistry, and conformation using methods such as high-resolution MS, MS/MS, NMR, chemical derivatization, crystallography, and synthesis.

Syntrophy

A metabolic relationship in which organisms depend on each other to process compounds or maintain favorable energetics. Syntrophy helps explain why isolated organisms may fail to grow or produce the same chemistry outside their communities.

Target engagement

Evidence that a molecule physically interacts with its proposed target in a relevant biological system at the achieved exposure. Target engagement is stronger than a prediction or isolated binding result but still does not guarantee clinical efficacy.

Tandem mass spectrometry (MS/MS)

A method in which selected ions are fragmented and the resulting product ions are measured. Fragmentation patterns provide structural information and enable molecular networking and library matching.

TGR5

A G-protein-coupled bile-acid receptor, also called GPBAR1. It responds to bile acids and participates in metabolic, inflammatory, and endocrine signaling in several tissues.

Toll-like receptor (TLR)

A family of innate immune receptors that recognize conserved microbial or damage-associated patterns. TLR signaling helps initiate defense but can also contribute to chronic inflammation when dysregulated.

Transcriptome

The set of RNA transcripts present in a cell, organism, community, or sample at a given time. The transcriptome reflects active regulation but is not a direct inventory of proteins or metabolites.

Transpeptidase

An enzyme that transfers a peptide or amino-acid group from one molecule to another. Transpeptidases participate in bacterial cell-wall assembly and can, in some contexts, create or modify cyclic and cross-linked peptides.

Tryptophan metabolite

A product derived from the amino acid tryptophan through host, microbial, dietary, or combined pathways. Examples include indole derivatives, kynurenine-pathway products, serotonin, and microbial aromatic metabolites.

Untargeted metabolomics

A broad survey of detectable molecular features without restricting the analysis to a predefined list. Untargeted studies are valuable for discovery but require careful annotation, controls, and confirmation.

Universal control layer

A future-facing systems concept in which context-rich chemical state measurements support feedback among sensing, interpretation, intervention, and learning. It is not a claim of one master controller or one universal healthy state.

Vagus nerve

The tenth cranial nerve and a major bidirectional communication route between visceral organs and the brainstem. It carries sensory and motor signals and participates in gut-brain communication, inflammation, and autonomic regulation.

Vertical gene transfer (VGT)

Transmission of genetic material from parent to offspring or from a cell to its descendants. Vertical inheritance contrasts with horizontal gene transfer between lineages.

Viable chemical candidate

In the finite-prior framework, a molecule or class that satisfies a defined set of requirements such as efficacy, safety, exposure, selectivity, manufacturability, and development feasibility. Viability is indication- and stage-specific.

Virome

The viruses present in a defined ecosystem. In human-associated environments, the virome includes phages, eukaryotic viruses, and genetic elements detected by sequencing. Counts and boundaries depend strongly on method.

Xenobiotic

A compound foreign to a particular organism or metabolic system, including drugs, pollutants, food additives, and environmental chemicals. Gut microbes and host enzymes can transform xenobiotics into more or less active products.

REFERENCE

Notes and Sources

These notes identify the principal evidence behind the narrative and flag places where the book moves from established findings to interpretation or hypothesis. They are not an exhaustive literature review. Full publication details are consolidated in the Selected Bibliography.

Author's Note, Preface, and Prologue

The four-level certainty framework is editorial rather than a scientific standard. It is designed to keep established finding, supported interpretation, author hypothesis, and commercial illustration visibly separate.

The working definitions of ghost genome, ghost molecule, ghost activity, chemical execution layer, observability, breakthrough aggressiveness, and universal control layer are the author's conceptual language. They are intended to generate testable distinctions, not to create new settled scientific categories.

Public Biosortia interviews and essays support the author's stated vision across therapeutics, agriculture, sustainability, remediation, and long-horizon human systems. These materials document the author's perspective; they are not independent validation of commercial or scientific claims.

The personal origin scenes draw on Ross Youngs's public interviews and recollection, including accounts of helping his father install medical-imaging equipment, studying environmental science and industrial engineering, building manufacturing companies, surviving a rapid market transition, and developing algal-harvesting technology that contributed to Biosortia's later platform. Dialogue is paraphrased unless explicitly identified as a quotation. The $6 million ARPA-E award and public career information are documented in Biosortia and third-party profiles.

Eroom's Law was named and analyzed by Scannell, Blanckley, Boldon, and Warrington in 2012. Their historical analysis described a major decline in inflation-adjusted drug-discovery productivity and offered multiple possible causes. The dilution framework and finite-prior interpretation are Ross Youngs's hypotheses, not a consensus explanation of pharmaceutical productivity.

The Muskegon harmful-algal-bloom example refers to public research on a prefractionated cyanobacterial library, molecular networking, product-ion queries, and cyanopeptide characterization. The book uses the work as evidence that large community collections can reveal molecular families; it does not imply that every feature has therapeutic value.

Chapter 1: The Gene Sweepstakes

The GeneSweep narrative is based on the National Human Genome Research Institute's historical exhibit and archive. Ewan Birney organized the wager at the Cold Spring Harbor genome meeting in May 2000, kept a betting book, and collected entries. More than 460 predictions were eventually submitted. The range extended from roughly twenty thousand to two hundred thousand genes. Lee Rowen's prediction of 25,947 was closest to the assessed count of 24,847 when the contest closed. Current protein-coding gene estimates are near twenty thousand and depend on annotation rules.

The Human Genome Project's completed euchromatic sequence was reported by the International Human Genome Sequencing Consortium. Early expectations exceeding one hundred thousand genes were common but not universal. The chapter's argument is not that the project failed; it is that gene count proved to be a poor measure of biological complexity.

The human and bacterial cell estimates draw on Sender, Fuchs, and Milo. Gut microbial gene catalogs are population references rather than the gene content of one person. The distinction between genetic capacity and chemical state is an interpretation developed across the book.

The strain and pangenome discussion draws on Peter et al. (2018), who reported 4,940 core and 2,856 variable open reading frames across 1,011 Saccharomyces cerevisiae isolates; Peris et al. (2023), who assembled ecological, geographic, genetic, and phylogenetic information for approximately 1,800 Saccharomyces strains across the genus; and Badet et al. (2020), whose 19-isolate Zymoseptoria tritici pangenome found approximately 60 percent core orthogroups and presence-absence variation in 16 of 39 syntenic secondary-metabolite clusters. These are population-level pangenome findings, not claims that every pair of strains differs by the same percentage.

Chapter 2: The Doctor Who Drank the Evidence

The H. pylori story draws on the Nobel Prize account, Barry Marshall's historical reviews, Robin Warren and Marshall's original clinical work, and later retrospective profiles. Warren observed curved bacteria in gastric biopsies. Marshall joined the project in 1981. Extended incubation over an Easter holiday contributed to successful culture. In the self-experiment, Marshall developed gastritis and the organism was recovered; the simplified claim that he produced a fully developed peptic ulcer is an overstatement.

The discovery did not show that all ulcers are infectious. Nonsteroidal anti-inflammatory drugs, acid physiology, host factors, strain variation, and other mechanisms remain important. The decisive achievement was the accumulated evidence that H. pylori causes chronic gastritis and is a major cause of peptic-ulcer disease and gastric cancer risk, enabling curative antimicrobial therapy in many patients.

The discussion of funding incentives draws on work by Alberts and colleagues, Azoulay and colleagues, Packalen and Bhattacharya, and programs created to support high-risk research. The evidence is more nuanced than a claim that peer review uniformly rejects novelty.

The Alzheimer's discussion is deliberately balanced. Amyloid biology is strongly supported, and lecanemab and donanemab showed modest slowing of decline in selected patients with early disease while carrying important risks and practical burdens. The point is that a valid mechanism can become an overly complete disease model.

Chapter 3: The Planet That Should Not Exist

Mayor and Queloz reported 51 Pegasi b in Nature in 1995 using radial-velocity measurements from the ELODIE spectrograph at Haute-Provence Observatory. The planet orbits a Sun-like star every 4.23 days and became the first confirmed exoplanet around such a star. It was surprising because prevailing formation models placed gas giants far from their stars. Later work established planetary migration and a wide diversity of system architectures.

The Galileo stories and the phrase E pur si muove are treated as cultural legends rather than verbatim transcripts. The Galileo Effect is the author's name for a recurring pattern, not a formal sociological law.

Woese and Fox used ribosomal RNA comparisons to reveal deep divisions among cellular life, including Archaea. The Great Plate Count Anomaly predates sequencing; culture-independent methods made the uncultured majority more legible.

The discussion of metabolomics dark matter draws on da Silva, Dorrestein, and Quinn and subsequent work in molecular networking and annotation. Unknown features include real molecules, known variants, adducts, fragments, contaminants, and artifacts. Unknown does not equal novel or important.

Chapter 4: The Molecule in the Freezer

The rapamycin narrative draws on historical accounts of the 1964 Canadian expedition to Rapa Nui, the isolation of Streptomyces hygroscopicus, Suren

Sehgal's work at Ayerst, the Montreal laboratory closure, and his preservation of cultures in a home freezer. Rapamycin was initially recognized for antifungal activity and later for antiproliferative and immunosuppressive effects. Sirolimus received U.S. approval in 1999.

Rapamycin binds FKBP12, and the resulting complex engages mTOR. The story illustrates induced-interface pharmacology and does not imply that all natural products act as molecular glues.

Lipinski's Rule of Five was designed as an empirical guide to oral absorption and permeation problems, not a universal law. Doak and colleagues and subsequent bRo5 research describe orally relevant chemical space beyond those conventional ranges. Natural origin does not imply human safety.

The master-key or change-key model is presented as an author hypothesis. Computationally predicted activity across several receptors is not proof of functional agonism or antagonism, in vivo exposure, target necessity, safety, or evolutionary purpose. This edition intentionally omits proprietary structures, target values, algorithms, and unpublished experimental details.

The discussion of biological signaling beyond 500 daltons uses cyclic dinucleotides as examples: cyclic di-AMP, 2'3'-cGAMP, and cyclic di-GMP. These are polar second messengers, not examples of conventional oral-drug permeability. Their relevance is that biological signaling has no hard 500-dalton boundary. The audio-recording-cost figure is an illustrative analogy rather than an audited economic series.

Chapter 5: The Twins Who Transferred a Metabolism

Ridaura and colleagues transplanted microbial communities from adult female twin pairs discordant for obesity into germ-free mice. Increased adiposity and associated metabolic phenotypes were transmissible under controlled conditions. Cohousing with mice harboring a lean co-twin's microbiota prevented the phenotype under particular dietary conditions, and rescue correlated with invasion by specific Bacteroidetes. These experiments establish causal capacity in mice, not a claim that human obesity is transferred as a single entity or that fecal transplantation is a general obesity therapy.

The host-microbial co-metabolism discussion draws on Nicholson, Holmes, Gordon, Wikoff, Dodd, Devlin, Fischbach, and a broad literature on bile acids, aromatic amino acids, short-chain fatty acids, and drug metabolism. The stronger cryptic-assembly hypothesis remains to be demonstrated and is presented with explicit experimental predictions.

Fan and colleagues reported RORDEP1 and RORDEP2, polypeptides produced by specific Ruminococcus torques strains, in culture supernatants and human plasma. Strain or gene abundance was inversely associated with adiposity in human datasets, and bacterial or recombinant-peptide interventions improved metabolic endpoints in rodents. These results are preclinical for efficacy and do not establish therapeutic benefit in humans.

Chapter 6: The Viruses in the Mucus

Barr and colleagues reported enrichment of phages in mucus and proposed the Bacteriophage Adherence to Mucus model. Experimental work supported interactions involving phage capsid immunoglobulin-like domains and mucin glycans and showed protection in model systems. The book does not treat BAM as a universal, fully established human immune mechanism.

The often-quoted 10^31 estimate refers to virus-like particles globally, not inside one person. Human gut phage catalogs contain extensive diversity, but sequence presence does not establish activity, host range, lysis, or chemical effect.

The proposal that phage lysis creates biologically important microbursts of specialized metabolites at mucosal surfaces is plausible but undercharacterized. It remains a research hypothesis requiring spatial and time-resolved measurement.

Chapter 7: The Lake That Would Not Fit in a Petri Dish

This chapter combines the author's firsthand experience with public descriptions of Industrial-Scale Microbiome Mining. It intentionally omits confidential source coordinates, operational parameters, proprietary yields, exact structures, and unpublished results.

The mass-balance example is illustrative. Actual recovery depends on concentration, partitioning, degradation, extraction efficiency, and many other losses. Scale does not establish biological value; it can make low-abundance chemistry experimentally accessible.

Public Biosortia materials document the 2009 ARPA-E award, the platform's evolution from algal harvesting, and research relationships with pharmaceutical and academic organizations. Company-reported performance claims should be treated as commercial illustrations until independently validated at molecule and portfolio level.

GNPS molecular networking, spatial metabolomics, stable-isotope tracing, NMR, authentic standards, and orthogonal biological assays form parts of the measurement stack described in the chapter.

Chapter 8: The Chip That Returned the Microbe Home

Nichols, Epstein, Lewis, and colleagues developed iChip methods that isolate individual microbial cells in diffusion chambers while exposing them to native environmental chemistry. Ling and colleagues reported teixobactin from an organism recovered through this approach. The phrase that teixobactin is the fish and the iChip is the rod is a common explanatory paraphrase, not a quotation attributed here to one source.

Teixobactin's initial report found no detectable resistance under the experimental conditions used. That result should not be generalized into a claim that resistance can never emerge.

antiSMASH and MIBiG are powerful tools for identifying and linking biosynthetic gene clusters. The chapter's corrective is that genetic capacity is not identical to final chemistry. Substrates, regulation, neighboring organisms, host metabolism, and nonenzymatic reactions can alter output.

Chapter 9: The Day AI Solved the Right Problem

AlphaFold 2's CASP14 performance and the 2021 Nature paper established a landmark in protein-structure prediction. AlphaFold 3, reported in Nature in 2024, extended prediction to joint complexes involving proteins, nucleic acids, small molecules, ions, and modified residues. These systems do not eliminate the need for experimental structural biology or prove efficacy, exposure, safety, or clinical value in drug discovery.

The discussion of data bias, model coverage, generation versus validation, and closed-loop learning draws on the machine-learning and drug-discovery literature. The phrase synthetic confidence is the author's description of mutually reinforcing predictions that have not been tested physically.

The finite-prior model comes from Youngs's 2025 preprint on finitude and dilution. It has not been prospectively validated as a general law. The proposed portfolio comparison is intended as a falsifiable test.

Breakthrough aggressiveness is used here as an innovation principle: challenge the architecture of the question while maintaining prospective benchmarks, negative-data visibility, independent confirmation, and conservative claim language.

The GDB-17 example comes from Ruddigkeit and colleagues, who enumerated 166.4 billion chemically plausible structures containing up to 17 heavy atoms. The count demonstrates combinatorial scale; an enumerated structure is not a synthesized, assayed, or biologically validated molecule.

Chapter 10: The Soil Microbe That Changed Global Health

Ōmura's Kitasato team collected and screened soil microorganisms, and a promising Streptomyces culture from the Kawana area was sent through the Kitasato-Merck collaboration. Campbell's group identified potent activity against parasitic worms, isolated avermectins, and worked with chemists who developed ivermectin. Mohammed Aziz helped champion human clinical development.

Merck committed in 1987 to donate Mectizan for river blindness for as long as needed, creating a public-private delivery program with international and local partners. By the mid-2020s, the program had delivered billions of treatments and contributed to elimination of river blindness or lymphatic filariasis in multiple countries. Exact totals change and should be checked near publication.

The business, patent, regulatory, and partnership discussion is strategic and general, not legal or investment advice. The clinical vignettes are research directions, not claims that microbiome-derived molecules will replace existing therapies.

The five horizons and universal-control-layer discussion expresses the author's long-range vision. Therapeutic, longevity, agricultural, environmental, and closed-loop-system applications will require separate evidence, regulatory pathways, and ethical frameworks. The chapter does not claim that one molecule, platform, or algorithm can control these systems.

The gut-brain-network discussion uses established categories of neural, endocrine, immune, vascular, and metabolic communication. A gut-restricted molecule can initiate distal effects without itself entering the brain, but each relay requires measured exposure and causal validation. The public-observatory discussion draws on official descriptions of the Human Genome Project, Human Microbiome Project and iHMP, The Cancer Genome Atlas, the NIH Molecular Libraries Program, LINCS, All of Us, HuBMAP, the NIH Common Fund Metabolomics program, and the NIH Bio Genesis Mission. These programs overlap and support different layers of biomedical measurement; none is represented here as a complete map of the ghost genome. Bio Genesis was announced in July 2026 and remains a forward-looking federal initiative.

Epilogue: And Yet It Moves

MiMeDB 2.0 reports 280 human/microbial co-metabolites in its current curated statistics. The count should be checked again immediately before publication because the database is actively maintained. It is not a uniform count of fully reconstructed causal pathways, native activity maps, or target-and-relay maps.

REFERENCE

Selected Bibliography

Ablasser, A., Goldeck, M., Cavlar, T., et al. "cGAS Produces a 2'-5'-Linked Cyclic Dinucleotide Second Messenger That Activates STING." Nature 498 (2013): 380-384. doi:10.1038/nature12306.

Abramson, J., Adler, J., Dunger, J., et al. "Accurate Structure Prediction of Biomolecular Interactions with AlphaFold 3." Nature 630 (2024): 493-500. doi:10.1038/s41586-024-07487-w.

Alberts, B., Kirschner, M. W., Tilghman, S., and Varmus, H. "Rescuing US Biomedical Research from Its Systemic Flaws." Proceedings of the National Academy of Sciences 111, no. 16 (2014): 5773-5777. doi:10.1073/pnas.1404402111.

All of Us Research Program. "The All of Us Researcher Workbench and Dataset." National Institutes of Health, accessed August 2026.

Almeida, A., Nayfach, S., Boland, M., et al. "A Unified Catalog of 204,938 Reference Genomes from the Human Gut Microbiome." Nature Biotechnology 39 (2021): 105-114. doi:10.1038/s41587-020-0603-3.

American Chemical Society. "Discovery of Ivermectin / Mectizan." National Historic Chemical Landmark, 2016.

Anighoro, A., Bajorath, J., and Rastelli, G. "Polypharmacology: Challenges and Opportunities in Drug Discovery." Journal of Medicinal Chemistry 57, no. 19 (2014): 7874-7887. doi:10.1021/jm5006463.

Atanasov, A. G., Zotchev, S. B., Dirsch, V. M., and Supuran, C. T. "Natural Products in Drug Discovery: Advances and Opportunities." Nature Reviews Drug Discovery 20 (2021): 200-216. doi:10.1038/s41573-020-00114-z.

Azoulay, P., Graff Zivin, J. S., and Manso, G. "Incentives and Creativity: Evidence from the Academic Life Sciences." RAND Journal of Economics 42, no. 3 (2011): 527-554. doi:10.1111/j.1756-2171.2011.00140.x.

Badet, T., Oggenfuss, U., Abraham, L., McDonald, B. A., and Croll, D. "A 19-Isolate Reference-Quality Global Pangenome for the Fungal Wheat Pathogen Zymoseptoria tritici." BMC Biology 18 (2020): 12. doi:10.1186/s12915-020-0744-3.

Barr, J. J., Auro, R., Furlan, M., et al. "Bacteriophage Adhering to Mucus Provide a Non-Host-Derived Immunity." Proceedings of the National Academy of Sciences 110, no. 26 (2013): 10771-10776. doi:10.1073/pnas.1305923110.

Birney, E., Rowen, L., and National Human Genome Research Institute. "Gene Sweepstakes (GeneSweep)." NHGRI History of Genomics Archive and Virtual Exhibit, updated 2026.

Blin, K., Shaw, S., Kloosterman, A. M., et al. "antiSMASH 7.0: New and Improved Predictions for Detection, Regulation, Chemical Structures and Visualisation." Nucleic Acids Research 51, no. W1 (2023): W46-W50. doi:10.1093/nar/gkad344.

Bogdanov, A., Salib, M. N., Chase, A. B., et al. "Small Molecule In Situ Resin Capture Provides a Compound-First Approach to Natural Product Discovery." Nature Communications 15 (2024): 5230. doi:10.1038/s41467-024-49367-x.

Camarillo-Guerrero, L. F., Almeida, A., Rangel-Pineros, G., Finn, R. D., and Lawley, T. D. "Massive Expansion of Human Gut Bacteriophage Diversity." Cell 184, no. 4 (2021): 1098-1109.e9. doi:10.1016/j.cell.2021.01.029.

Campbell, W. C., Fisher, M. H., Stapley, E. O., Albers-Schonberg, G., and Jacob, T. A. "Ivermectin: A Potent New Antiparasitic Agent." Science 221, no. 4613 (1983): 823-828. doi:10.1126/science.6308762.

Christensen, C. M. The Innovator's Dilemma: When New Technologies Cause Great Firms to Fail. Boston: Harvard Business School Press, 1997.

Cimermancic, P., Medema, M. H., Claesen, J., et al. "Insights into Secondary Metabolism from a Global Analysis of Prokaryotic Biosynthetic Gene Clusters." Cell 158, no. 2 (2014): 412-421. doi:10.1016/j.cell.2014.06.034.

Cohen, L. J., Esterhazy, D., Kim, S.-H., et al. "Commensal Bacteria Make GPCR Ligands That Mimic Human Signalling Molecules." Nature 549 (2017): 48-53. doi:10.1038/nature23874.

Corrigan, R. M., and Gründling, A. "Cyclic di-AMP: Another Second Messenger Enters the Fray." Nature Reviews Microbiology 11 (2013): 513-524. doi:10.1038/nrmicro3069.

Crick, F. H. C. "On Protein Synthesis." Symposia of the Society for Experimental Biology 12 (1958): 138-163.

da Silva, R. R., Dorrestein, P. C., and Quinn, R. A. "Illuminating the Dark Matter in Metabolomics." Proceedings of the National Academy of Sciences 112, no. 41 (2015): 12549-12550. doi:10.1073/pnas.1516878112.

Davies, J. "Are Antibiotics Naturally Antibiotics?" Journal of Industrial Microbiology and Biotechnology 33 (2006): 496-499. doi:10.1007/s10295-006-0112-5.

Devlin, A. S., and Fischbach, M. A. "A Biosynthetic Pathway for a Prominent Class of Microbiota-Derived Bile Acids." Nature Chemical Biology 11 (2015): 685-690. doi:10.1038/nchembio.1864.

Doak, B. C., Over, B., Giordanetto, F., and Kihlberg, J. "Oral Druggable Space beyond the Rule of 5: Insights from Drugs and Clinical Candidates." Chemistry & Biology 21, no. 9 (2014): 1115-1142. doi:10.1016/j.chembiol.2014.08.013.

Dodd, D., Spitzer, M. H., Van Treuren, W., et al. "A Gut Bacterial Pathway Metabolizes Aromatic Amino Acids into Nine Circulating Metabolites." Nature 551 (2017): 648-652. doi:10.1038/nature24661.

Donia, M. S., and Fischbach, M. A. "Small Molecules from the Human Microbiota." Science 349, no. 6246 (2015): 1254766. doi:10.1126/science.1254766.

Donia, M. S., Cimermancic, P., Schulze, C. J., et al. "A Systematic Analysis of Biosynthetic Gene Clusters in the Human Microbiome Reveals a Common Family of Antibiotics." Cell 158, no. 6 (2014): 1402-1414. doi:10.1016/j.cell.2014.08.032.

Dorrestein, P. C., Mazmanian, S. K., and Knight, R. "Finding the Missing Links among Metabolites, Microbes, and the Host." Immunity 40, no. 6 (2014): 824-832. doi:10.1016/j.immuni.2014.05.015.

Fan, Y., Lyu, L., Vazquez-Uribe, R., et al. "Polypeptides Synthesized by Common Bacteria in the Human Gut Improve Rodent Metabolism." Nature Microbiology 10 (2025): 1918-1939. doi:10.1038/s41564-025-02064-x.

Filella-Merce, I., Molina, A., Diaz, L., et al. "Optimizing Drug Design by Merging Generative AI with a Physics-Based Active Learning Framework." Communications Chemistry 8 (2025): 238. doi:10.1038/s42004-025-01635-7.

Haiser, H. J., Gootenberg, D. B., Chatman, K., Sirasani, G., Balskus, E. P., and Turnbaugh, P. J. "Predicting and Manipulating Cardiac Drug Inactivation by the Human Gut Bacterium Eggerthella lenta." Science 341, no. 6143 (2013): 295-298. doi:10.1126/science.1235872.

Hobby, G., Clark, R., and Woywodt, A. "A Treasure from a Barren Island: The Discovery of Rapamycin." Clinical Kidney Journal 15, no. 10 (2022): 1971-1972. doi:10.1093/ckj/sfac116.

Hopkins, A. L. "Network Pharmacology: The Next Paradigm in Drug Discovery." Nature Chemical Biology 4 (2008): 682-690. doi:10.1038/nchembio.118.

HuBMAP Consortium. "The Human BioMolecular Atlas Program and Human Reference Atlas." National Institutes of Health, accessed August 2026.

Human Microbiome Project Consortium. "Structure, Function and Diversity of the Healthy Human Microbiome." Nature 486 (2012): 207-214. doi:10.1038/nature11234.

Human Microbiome Project Data Coordination Center. "Integrative Human Microbiome Project Resources." National Institutes of Health, accessed August 2026.

International Human Genome Sequencing Consortium. "Finishing the Euchromatic Sequence of the Human Genome." Nature 431 (2004): 931-945. doi:10.1038/nature03001.

Jenal, U., Reinders, A., and Lori, C. "Cyclic di-GMP: Second Messenger Extraordinaire." Nature Reviews Microbiology 15 (2017): 271-284. doi:10.1038/nrmicro.2016.190.

Jumper, J., Evans, R., Pritzel, A., et al. "Highly Accurate Protein Structure Prediction with AlphaFold." Nature 596 (2021): 583-589. doi:10.1038/s41586-021-03819-2.

Koppel, N., Rekdal, V. M., and Balskus, E. P. "Chemical Transformation of Xenobiotics by the Human Gut Microbiota." Science 356, no. 6344 (2017): eaag2770. doi:10.1126/science.aag2770.

Kretschmer, F., Seipp, J., Ludwig, M., et al. "Coverage Bias in Small Molecule Machine Learning." Nature Communications 16 (2025): 554. doi:10.1038/s41467-024-55462-w.

Kruger, R., Oler, E., Saha, S., et al. "MiMeDB 2.0: The Human Microbial Metabolome Database for 2026." Nucleic Acids Research 54, no. D1 (2026): D623-D634. doi:10.1093/nar/gkaf1272.

Kumar, D. K. V., Choi, S. H., Washicosky, K. J., et al. "Amyloid-beta Peptide Protects against Microbial Infection in Mouse and Worm Models of Alzheimer's Disease." Science Translational Medicine 8, no. 340 (2016): 340ra72. doi:10.1126/scitranslmed.aaf1059.

Li, J., Jia, H., Cai, X., et al. "An Integrated Catalog of Reference Genes in the Human Gut Microbiome." Nature Biotechnology 32 (2014): 834-841. doi:10.1038/nbt.2942.

Ling, L. L., Schneider, T., Peoples, A. J., et al. "A New Antibiotic Kills Pathogens without Detectable Resistance." Nature 517 (2015): 455-459. doi:10.1038/nature14098.

Lipinski, C. A., Lombardo, F., Dominy, B. W., and Feeney, P. J. "Experimental and Computational Approaches to Estimate Solubility and Permeability in Drug Discovery and Development Settings." Advanced Drug Delivery Reviews 23, no. 1-3 (1997): 3-25. doi:10.1016/S0169-409X(96)00423-1.

Lovering, F., Bikker, J., and Humblet, C. "Escape from Flatland: Increasing Saturation as an Approach to Improving Clinical Success." Journal of Medicinal Chemistry 52, no. 21 (2009): 6752-6756. doi:10.1021/jm901241e.

Marshall, B. J. "The Campylobacter Pylori Story." Scandinavian Journal of Gastroenterology Supplement 146 (1988): 58-66, and later historical reviews.

Mayor, M., and Queloz, D. "A Jupiter-Mass Companion to a Solar-Type Star." Nature 378 (1995): 355-359. doi:10.1038/378355a0.

Mectizan Donation Program. "History of the Program." The Task Force for Global Health and Merck, accessed 2026.

Medema, M. H., Blin, K., Cimermancic, P., et al. "antiSMASH: Rapid Identification, Annotation and Analysis of Secondary Metabolite Biosynthesis Gene Clusters." Nucleic Acids Research 39, suppl. 2 (2011): W339-W346. doi:10.1093/nar/gkr466.

National Aeronautics and Space Administration. "Nobel Winners Changed Our Understanding with Exoplanet Discovery: 51 Pegasi b." NASA Science.

National Cancer Institute. "The Cancer Genome Atlas Program." National Institutes of Health, accessed August 2026.

National Institutes of Health. "Bio Genesis Mission." July 22, 2026.

National Institutes of Health. "The Nobel Prize in Physiology or Medicine 2005: Barry J. Marshall and J. Robin Warren." NobelPrize.org.

Newman, D. J., and Cragg, G. M. "Natural Products as Sources of New Drugs over the Nearly Four Decades from 01/1981 to 09/2019." Journal of Natural Products 83, no. 3 (2020): 770-803. doi:10.1021/acs.jnatprod.9b01285.

Nichols, D., Cahoon, N., Trakhtenberg, E. M., et al. "Use of Ichip for High-Throughput In Situ Cultivation of 'Uncultivable' Microbial Species." Applied and Environmental Microbiology 76, no. 8 (2010): 2445-2450. doi:10.1128/AEM.01754-09.

Nicholson, J. K., Holmes, E., Kinross, J., Burcelin, R., Gibson, G., Jia, W., and Pettersson, S. "Host-Gut Microbiota Metabolic Interactions." Science 336, no. 6086 (2012): 1262-1267. doi:10.1126/science.1223813.

NIH Common Fund. "Library of Integrated Network-Based Cellular Signatures (LINCS)." National Institutes of Health, accessed August 2026.

NIH Common Fund. "Metabolomics Program and Metabolomics Workbench Resources." National Institutes of Health, accessed August 2026.

NIH Common Fund. "Molecular Libraries and Imaging Program." National Institutes of Health, accessed August 2026.

Omura, S., and Crump, A. "The Life and Times of Ivermectin - A Success Story." Nature Reviews Microbiology 2 (2004): 984-989. doi:10.1038/nrmicro1048.

Packalen, M., and Bhattacharya, J. "NIH Funding and the Pursuit of Edge Science." Proceedings of the National Academy of Sciences 117, no. 22 (2020): 12011-12016. doi:10.1073/pnas.1910160117.

Peris, D., Ubbelohde, E. J., Kuang, M. C., et al. "Macroevolutionary Diversity of Traits and Genomes in the Model Yeast Genus Saccharomyces." Nature Communications 14 (2023): 690. doi:10.1038/s41467-023-36139-2.

Peter, J., De Chiara, M., Friedrich, A., et al. "Genome Evolution across 1,011 Saccharomyces cerevisiae Isolates." Nature 556 (2018): 339-344. doi:10.1038/s41586-018-0030-5.

Qin, J., Li, R., Raes, J., et al. "A Human Gut Microbial Gene Catalogue Established by Metagenomic Sequencing." Nature 464 (2010): 59-65. doi:10.1038/nature08821.

Ridaura, V. K., Faith, J. J., Rey, F. E., et al. "Gut Microbiota from Twins Discordant for Obesity Modulate Metabolism in Mice." Science 341, no. 6150 (2013): 1241214. doi:10.1126/science.1241214.

Ruddigkeit, L., van Deursen, R., Blum, L. C., and Reymond, J.-L. "Enumeration of 166 Billion Organic Small Molecules in the Chemical Universe Database GDB-17." Journal of Chemical Information and Modeling 52, no. 11 (2012): 2864-2875. doi:10.1021/ci300415d.

Salminen, S., Collado, M. C., Endo, A., et al. "The International Scientific Association of Probiotics and Prebiotics Consensus Statement on the Definition and Scope of Postbiotics." Nature Reviews Gastroenterology & Hepatology 18 (2021): 649-667. doi:10.1038/s41575-021-00440-6.

Scannell, J. W., Blanckley, A., Boldon, H., and Warrington, B. "Diagnosing the Decline in Pharmaceutical R&D Efficiency." Nature Reviews Drug Discovery 11 (2012): 191-200. doi:10.1038/nrd3681.

Schneider, P., Walters, W. P., Plowright, A. T., et al. "Rethinking Drug Design in the Artificial Intelligence Era." Nature Reviews Drug Discovery 19 (2020): 353-364. doi:10.1038/s41573-019-0050-3.

Sender, R., Fuchs, S., and Milo, R. "Revised Estimates for the Number of Human and Bacteria Cells in the Body." PLoS Biology 14, no. 8 (2016): e1002533. doi:10.1371/journal.pbio.1002533.

Sims, J. R., Zimmer, J. A., Evans, C. D., et al. "Donanemab in Early Symptomatic Alzheimer Disease: The TRAILBLAZER-ALZ 2 Randomized Clinical Trial." JAMA 330, no. 6 (2023): 512-527. doi:10.1001/jama.2023.13239.

Soscia, S. J., Kirby, J. E., Washicosky, K. J., et al. "The Alzheimer's Disease-Associated Amyloid beta-Protein Is an Antimicrobial Peptide." PLoS ONE 5, no. 3 (2010): e9505. doi:10.1371/journal.pone.0009505.

Swinney, D. C., and Anthony, J. "How Were New Medicines Discovered?" Nature Reviews Drug Discovery 10 (2011): 507-519. doi:10.1038/nrd3480.

Tang, L. J. "Analysis of China-to-West Pharmaceutical Licensing Deals in 2024." Nature Reviews Drug Discovery 24 (2025): 411-412. doi:10.1038/d41573-025-00068-0.

Terlouw, B. R., Blin, K., Navarro-Munoz, J. C., et al. "MIBiG 3.0: A Community-Driven Effort to Annotate Experimentally Validated Biosynthetic Gene Clusters." Nucleic Acids Research 51, no. D1 (2023): D603-D610. doi:10.1093/nar/gkac1049.

Udenigwe, C. C., and Aluko, R. E. "Food Protein-Derived Bioactive Peptides: Production, Processing, and Potential Health Benefits." Journal of Food Science 77, no. 1 (2012): R11-R24. doi:10.1111/j.1750-3841.2011.02455.x.

Vamathevan, J., Clark, D., Czodrowski, P., et al. "Applications of Machine Learning in Drug Discovery and Development." Nature Reviews Drug Discovery 18 (2019): 463-477. doi:10.1038/s41573-019-0024-5.

van Dyck, C. H., Swanson, C. J., Aisen, P., et al. "Lecanemab in Early Alzheimer's Disease." New England Journal of Medicine 388 (2023): 9-21. doi:10.1056/NEJMoa2212948.

Wallace, B. D., Wang, H., Lane, K. T., et al. "Alleviating Cancer Drug Toxicity by Inhibiting a Bacterial Enzyme." Science 330, no. 6005 (2010): 831-835. doi:10.1126/science.1191175.

Wang, M., Carver, J. J., Phelan, V. V., et al. "Sharing and Community Curation of Mass Spectrometry Data with Global Natural Products Social Molecular Networking." Nature Biotechnology 34 (2016): 828-837. doi:10.1038/nbt.3597.

Wikoff, W. R., Anfora, A. T., Liu, J., et al. "Metabolomics Analysis Reveals Large Effects of Gut Microflora on Mammalian Blood Metabolites." Proceedings of the National Academy of Sciences 106, no. 10 (2009): 3698-3703. doi:10.1073/pnas.0812874106.

Woese, C. R., and Fox, G. E. "Phylogenetic Structure of the Prokaryotic Domain: The Primary Kingdoms." Proceedings of the National Academy of Sciences 74, no. 11 (1977): 5088-5090. doi:10.1073/pnas.74.11.5088.

Wu, K., Wang, Y., Liu, W., et al. "TamGen: Drug Design with Target-Aware Molecule Generation through a Chemical Language Model." Nature Communications 15 (2024): 9360. doi:10.1038/s41467-024-53632-4.

Xia, R., Ahn, L., Burkhauser, M., Youngs, R., and Bertin, M. J. "LC-MS/MS Strategy Using Substructure Searching for the Annotation of Cyanopeptide Classes: Implications for New Compound Discovery and Environmental Monitoring." Environmental Science & Technology 60, no. 26 (2026): 18545-18554. doi:10.1021/acs.est.6c05607.

Youngs, R. O. "Industrial Scale Microbiome Mining with Biosortia's Ross Youngs." Interview by Infinite Frontiers, June 11, 2024.

Youngs, R. O. "Microbiome Metabolic Fingerprinting: A Critical Control Layer for Adaptive, Closed-Loop Biological Systems." Zenodo preprint, 2025. doi:10.5281/zenodo.17727788.

Youngs, R. O. "Mining the Microbiome" and related public interviews and presentations. Biosortia videos and podcasts, accessed August 2026.

Youngs, R. O. "Reversing Eroom's Law: Finitude and Dilution in Small-Molecule Discovery." bioRxiv preprint, 2025. doi:10.1101/2025.10.29.685434.

Youngs, R. O. "The Beauty of Microbial Natural Products: Evolutionary Polypharmacology and the Difference between Coherent Multi-Target Activity and Promiscuity." Zenodo white paper, 2025. doi:10.5281/zenodo.17445771.

Youngs, R. O. "The Evolutionary Advantage: Why Microbial Chemistry Wins in Drug Discovery and Approval." Zenodo white paper, 2025. doi:10.5281/zenodo.17445695.

Zimmermann, M., Zimmermann-Kogadeeva, M., Wegmann, R., and Goodman, A. L. "Mapping Human Microbiome Drug Metabolism by Gut Bacteria and Their Genes." Nature 570 (2019): 462-467. doi:10.1038/s41586-019-1291-3.

www.ingramcontent.com/pod-product-compliance
Lightning Source LLC
LaVergne TN
LVHW022024051126
851858LV00003B/32
* 9 7 9 8 9 9 7 3 6 0 5 2 8 *